RESONANCE
Applications in Physical Science

RESONANCE

Applications in Physical Science

Michael M. Woolfson

University of York, UK

Imperial College Press

Published by

Imperial College Press
57 Shelton Street
Covent Garden
London WC2H 9HE

Distributed by

World Scientific Publishing Co. Pte. Ltd.
5 Toh Tuck Link, Singapore 596224
USA office: 27 Warren Street, Suite 401-402, Hackensack, NJ 07601
UK office: 57 Shelton Street, Covent Garden, London WC2H 9HE

Library of Congress Cataloging-in-Publication Data
Woolfson, Michael M. (Michael Mark)
   Resonance : applications in physical science / Michael Mark Woolfson, University of York, UK.
      pages cm
   Includes bibliographical references and index.
   ISBN 978-1-78326-538-1 (hardcover : alk. paper) -- ISBN 978-1-78326-539-8 (pbk. : alk. paper)
   1. Resonance. I. Title.
   QC241.W66 2014
   530.4'16--dc23
                                                   2014035510

British Library Cataloguing-in-Publication Data
A catalogue record for this book is available from the British Library.

Cover image: St Paul's Cathedral, London, viewed from south bank end of Millennium Footbridge, near Tate Modern.
Source: Wikipedia Commons.

Typeset by Stallion Press
Email: enquiries@stallionpress.com

Contents

Introduction

Many centuries ago, an intelligent individual could master a large proportion of the prevailing knowledge of the time and there were no accepted boundaries in science, as we know them today. Thus, Isaac Newton was not only a mathematician and physicist but also spent much of his later life in the practice of alchemy, a kind of chemistry not held in high regard, even in those days. Robert Hooke, a contemporary of Newton, had even wider interests, covering gravitation, astronomy, biology, palaeontology and the design of timepieces.

With the explosive expansion of scientific knowledge, particularly in the last few decades, it has become increasingly difficult to keep up-to-date even in one area of science — be it physics, chemistry, biology or any other well-defined subject area. Indeed, it is a well-rounded physicist who can understand even 50% of the titles in the publication *Physics Abstracts*, let alone the contents of the papers to which they refer.

It has been traditional to structure university science lecture courses in terms of mainstream topics (e.g. physical optics, structure and bonding, atomic physics, spectroscopy etc.) and to many students the links between different strands of their course never become clear. Not only do they not see the essential unity of science, but also some even lack a sense of unity in their own specialist area of study.

It is clearly impossible to return to the era of the universal scientist, but what *is* possible is to explore some topics which act like weft in binding together different strands within a single subject area or even, to some extent, different subject areas. Various branches of mathematics serve this purpose: as an example, matrices can be used to analyse complex lens systems; to represent the Lorentz transformation in relativity theory; to express electron spin in quantum mechanics; and to study both vibrating systems and biological populations.

At the level of the trained scientist, the appreciation of these cross-links can give insights that lead to the better practice of science. X-ray crystallography is an example of a topic that binds together vast tracts of science such as mathematics, physics, chemistry, biology, geology and mineralogy — and there may be others. The American pure mathematician Herbert Hauptman (1917–2011), who became a theoretical crystallographer, eventually won a Nobel Prize in Chemistry!

Another of these linking topics is *resonance*, which occurs in various guises: in everyday life; in artistic contexts such as music; and in many forms in the physical sciences. The purpose of this textbook is to explore a variety of scientific areas in which resonance occurs, ranging from concerns about the safety of bridges to the confirmation of a prediction of Einstein's Theory of General Relativity.

Interspersed within the chapters are *exercises*, which are usually simple numerical applications of the immediately preceding material and are designed to instil a feeling for the magnitudes of the quantities of interest. At the end of chapters there are *problems*, which are more demanding than the exercises and whose purpose is to test the student's understanding of the material within each chapter. A solutions section provides a check on the student's success (or otherwise) in completing these tests, but where the student has difficulties, they also provide a guide that reinforces the teaching role of the chapter.

Before the advent of the computer age there were two branches of every science: *experimental* and *theoretical*. To these we must now add *computational*; there are many areas of science where both experimental and theoretical solutions to problems are impossible and the only solution is through computation. An outstanding example is the solution of many-body problems in astronomy;

there are some theoretical solutions for special three-body problems (see Section 4.4.4) but, in general, the motion of more than two bodies under their mutual gravitational forces requires a computational approach. Appendices II and III give programs to solve astronomical many-body problems, which are required for the complete solution of some of these problems.[1] They are in very basic FORTRAN77 code, but for those unfamiliar with FORTRAN the program will guide them in inserting the correct data, if they have access to a compiler.

A useful ability for most modern scientists is to be able quickly to write short programs to carry out straightforward calculations that would be tedious any other way. There are problems in this text where the value of a function of a single variable is required for a large number of values of the variable and simple programs consisting of few lines of code enable this to be done quickly. By outputting data to a file, a graphical solution can then be found with the use of a graphics package.

---

[1] Copies of these programs are available for free download from:
http://www.worldscientific.com/worldscibooks/10.1142/p963#t=suppl

Chapter 1

Simple Harmonic Motion, Damping and Resonance

1.1 Simple Harmonic Motion

In 1582 the Italian mathematician and astronomer Galileo Galilei (1564–1642) observed the swinging motion of a chandelier in Pisa Cathedral. He noticed that the period of its swing, which it is said that he timed using his pulse, did not change as the amplitude of the swing died down. In 1602 he carried out a series of experiments on the properties of a pendulum and later, towards the end of his life, he conceived the idea of using a pendulum to regulate a mechanical clock. Such clocks were made after he died and were the most precise timepieces available until comparatively recently.

1.1.1 *A mass on a vertical spring*

The type of motion performed by the pendulum is known as *simple harmonic motion*. However, although Galileo observed no variation of period with the amplitude of the swing there *is* a dependence that would only be measurable for moderately large swings with instruments better than a pulse. For that reason we shall consider another system that gives simple harmonic motion: a mass, m, attached to the end of a light vertical spring, as shown in Figure 1.1.

The downward force on the spring due to the mass is its weight, mg, (where g is the acceleration due to gravity) and this stretches the

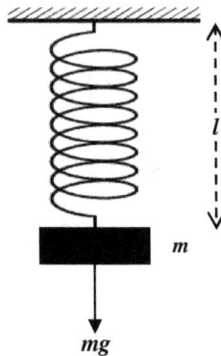

Figure 1.1 A mass on a spring.

spring until the upward force due to the spring balances the weight and the mass is then stationary at a point of equilibrium. A normal spring, not under excessive strain, satisfies Hooke's law that the force required either to extend or compress it will be proportional to the extension or compression. This means that if we displace the mass in Figure 1.1 by a distance x, which can be either positive or negative, the net force on it will be

$$F = -\kappa x, \tag{1.1}$$

in which κ, the *stiffness* of the spring, is the force per unit displacement and the negative sign indicates that the force is in the opposite direction to the displacement. This force will give an acceleration of the mass so we can write

$$m\frac{d^2x}{dt^2} = -\kappa x$$

or

$$\frac{d^2x}{dt^2} = -\frac{\kappa}{m}x = -\omega^2 x, \tag{1.2}$$

in which we have introduced $\omega = \sqrt{\kappa/m}$.

The general solution to the differential equation (1.2) is

$$x = a_1 \cos(\omega t) + a_2 \sin(\omega t), \tag{1.3a}$$

which can be reformulated as

$$x = A\cos(\omega t + \phi) \tag{1.3b}$$

where

$$A = \sqrt{a_1^2 + a_2^2} \quad \text{and} \quad \tan(\phi) = \frac{-a_2}{a_1}. \tag{1.3c}$$

The ambiguity in the value of ϕ from (1.3c) is resolved by noting that $\sin(\phi)$ has the sign of $-a_2$ and $\cos(\phi)$ has the sign of a_1. Equation (1.3b) defines simple harmonic motion: the quantity A, the maximum displacement, is the *amplitude* of the motion and ϕ is called the *phase angle*.

A way of representing simple harmonic motion is as the projection on a diameter of a uniform circular motion, as shown in Figure 1.2. The circular motion begins at point P_0 when $t = 0$ and moves round the circumference with angular speed ω. After time t the point P is reached and the projected distance OQ is $A\cos(\omega t + \phi)$. The *frequency*, n, of the simple harmonic motion is the number of oscillations per second, the unit of which is the *hertz* (Hz) corresponding to 1 oscillation per second. Thus a frequency of 10 Hz indicates 10 oscillations per second. Frequency is given by

$$n = \frac{\omega}{2\pi} \tag{1.4}$$

and ω, which is proportional to n is called the *angular frequency*. The *period* of the simple harmonic motion, P, the time for one complete

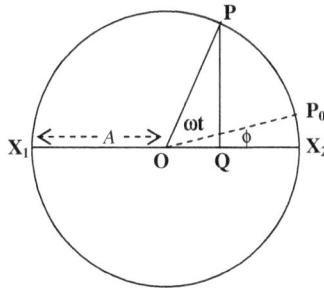

Figure 1.2 Simple harmonic motion interpreted as projected circular motion.

oscillation is

$$P = \frac{1}{n} = \frac{2\pi}{\omega}.$$ (1.5)

Exercise 1.1   A mass of 1 kg at the end of a spring experiences a force of 1 N for a displacement of 1 cm. If the mass is displaced from its equilibrium position and released then what is its period of oscillation?

1.1.2   *The simple pendulum*

As long as the force is proportional to the displacement of the spring, the analysis we used gives the correct solution for the motion. However, for the simple pendulum, illustrated in Figure 1.3 as a small mass, m, at the end of a cord of zero mass and length l, the analysis breaks down for large amplitudes of swing.

The simple pendulum is best treated as a rotating system where the mass rotating about the point O at distance l has a *moment of inertia* ml^2 and the torque acting on the system is Fl, where F is the component of the weight perpendicular to the cord and is $mg\sin(\theta)$. Now, equating torque to moment of inertia × angular acceleration we have

$$ml^2\frac{d^2\theta}{dt^2} = -mgl\sin(\theta).$$ (1.6)

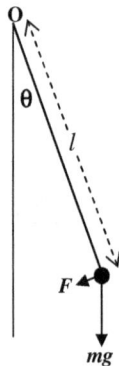

Figure 1.3 The simple pendulum.

If θ is small then $\sin(\theta) \approx \theta$ and Equation (1.6) can be written as

$$\frac{d^2\theta}{dt^2} = -\frac{g}{l}\theta$$

with solution

$$\theta = \Theta\cos(\omega t + \phi) \tag{1.7}$$

where Θ is the *angular amplitude*, expressed in *radians*,[1] and $\omega = \sqrt{g/l}$. The period of the pendulum, which Galileo measured, is, from Equation (1.7)

$$P = \frac{2\pi}{\omega} = 2\pi\sqrt{\frac{l}{g}}. \tag{1.8}$$

The solution for the simple pendulum depends on the approximation $\sin(\theta) \approx \theta$, which breaks down for large θ. For larger amplitudes of swing the period is given by

$$P = 2\pi\sqrt{\frac{l}{g}}\left(1 + \frac{1}{16}\Theta^2 + \frac{11}{3072}\Theta^4 + \cdots\right). \tag{1.9}$$

The brackets on the right-hand side contain the first three terms of an infinite series but, for modest amplitudes of swing, the first two terms are usually sufficient. For an amplitude of $\Theta = 0.2$ radians (about $11.5°$) the second term is 0.0025 and the third term 5.73×10^{-6}.

Exercise 1.2 What length of a simple pendulum will give a period of $1\,\mathrm{s}$ if the angular amplitude of swing is $25°$ $(g = 9.8\,\mathrm{m\cdot s^{-2}})$.

1.1.3 *The energy of simple harmonic motion*

At any time the energy of motion of the system can be divided into two parts: potential energy and kinetic energy. The *potential energy*, Ω, is the work done to move the mass from the equilibrium

---

[1] A radian is the angle subtended at the centre of a circle by an arc on its circumference equal in length to the radius. It is a dimensionless quantity equivalent to $180/\pi = 57.296°$.

position ($x = 0$) to its instantaneous location. At a position with coordinate x the force exerted by the spring is κx and the work done to stretch it by a further distance dx is

$$d\Omega = \kappa x dx.$$

Hence the total work done to stretch the spring from $x = 0$ to $x = X$, which is the potential energy (i.e. the energy stored in the stretched or compressed spring) is

$$\Omega = \int_0^X \kappa x dx = \frac{1}{2}\kappa X^2 = \frac{1}{2}\kappa A^2 \cos^2(\omega t_X + \phi) \qquad (1.10)$$

where t_X is the time at which $x = X$.

The *kinetic energy*, the energy of motion of the mass, involves the speed at the point X, which is given by

$$V_X = \left(\frac{dx}{dt}\right)_{x=X} = -A\omega \sin(\omega t_X + \phi)$$

so that the kinetic energy of the mass is

$$K = \frac{1}{2}mV_X^2 = \frac{1}{2}m\omega^2 A^2 \sin^2(\omega t_X + \phi) = \frac{1}{2}\kappa A^2 \sin^2(\omega t_X + \phi).$$
$$(1.11)$$

The total energy associated with the simple harmonic motion is therefore

$$E = \Omega + K = \frac{1}{2}\kappa A^2\{\cos^2(\omega t_X + \phi) + \sin^2(\omega t_X + \phi)\} = \frac{1}{2}\kappa A^2.$$
$$(1.12)$$

This dependence of the total energy on the square of the amplitude is common to all kinds of simple harmonic motion, with the factor κ coming from the type and characteristics of the system.

Exercise 1.3 A mass of 0.5 kg is suspended from a spring for which the force per unit displacement of 60 N m^{-1}. If it is oscillating with amplitude 5 cm then what is its speed when it passes through the equilibrium point?

1.2   Damped Simple Harmonic Motion

The periodic solutions for simple harmonic motion, Equations (1.3b) and (1.7), have amplitudes that are time-invariant so that they continue indefinitely. When Galileo observed the swinging chandelier in Pisa Cathedral the amplitude of the swings gradually died down, which was due to *damping* caused mainly by the resistance of the air, leading to a loss of energy of the system manifested by a reduction in the amplitude of the motion.

For many types of resistance, especially that depending on viscosity, the resisting force is proportional to the speed with which a body moves through the resisting medium. In place of the force acting on the mass given in Equation (1.1) we now have

$$F = -\kappa x - f\frac{dx}{dt}. \tag{1.13}$$

The second term on the right-hand side is a resisting force proportional to the speed and acting in a direction opposed to the motion. The resulting differential equation is

$$\frac{d^2x}{dt^2} + \frac{f}{m}\frac{dx}{dt} + \omega^2 x = 0 \tag{1.14}$$

in which, as before, $\omega^2 = \kappa/m$.

A solution to Equation (1.14) can be found of the form

$$x = A\exp(\alpha t) \tag{1.15a}$$

giving

$$\frac{dx}{dt} = A\alpha\exp(\alpha t) \tag{1.15b}$$

and

$$\frac{d^2x}{dt^2} = A\alpha^2\exp(\alpha t). \tag{1.15c}$$

Inserting (1.15a), (1.15b) and (1.15c) into Equation (1.14) and dividing throughout by $A\exp(\alpha t)$ gives

$$\alpha^2 + \frac{f}{m}\alpha + \omega^2 = 0,$$

the solution of which is

$$\alpha = -\frac{f}{2m} \pm \sqrt{\frac{f^2}{4m^2} - \omega^2}. \tag{1.16}$$

This gives the most general solution of Equation (1.14) as

$$x = \exp\left(-\frac{f}{2m}t\right)\left\{A\exp\left(\sqrt{\frac{f}{4m^2} - \omega^2}t\right)\right.$$

$$\left. + B\exp\left(-\sqrt{\frac{f^2}{4m^2} - \omega^2}t\right)\right\}. \tag{1.17}$$

The form of this relationship depends on the nature of the expression under the square-root sign and we now consider three possibilities.

1.2.1 *Light damping*

In this case f is small enough for the quantity under the square-root sign to remain negative and we write

$$\frac{f^2}{4m^2} - \omega^2 = -\omega_l^2, \tag{1.18}$$

which gives

$$x = \exp\left(-\frac{f}{2m}t\right)\left\{A\exp(i\omega_l t) + B\exp(-i\omega_l t)\right\}. \tag{1.19}$$

The solution must be real, which can be achieved by choosing appropriate values of A and B. We take

$$A = \frac{1}{2}C\exp(i\phi) \quad \text{and} \quad B = \frac{1}{2}C\exp(-i\phi)$$

to give

$$x = \exp\left(-\frac{f}{2m}t\right)\left[\frac{1}{2}C\exp\{i(\omega_l t + \phi)\} + \frac{1}{2}C\exp\{-i(\omega_l t + \phi)\}\right]$$

$$= C\exp\left(-\frac{f}{2m}t\right)\cos(\omega_l t + \phi). \tag{1.20}$$

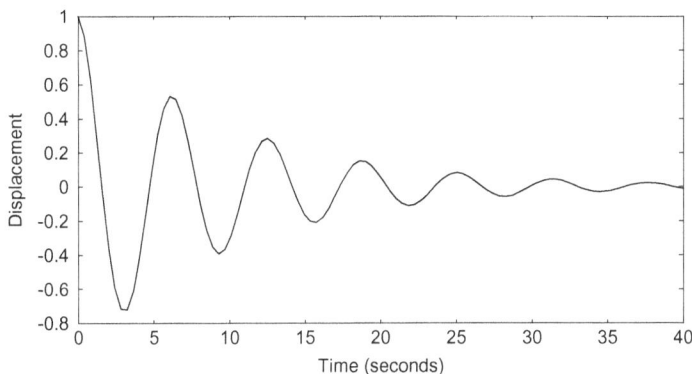

Figure 1.4 An example of a lightly damped simple harmonic oscillator.

The form of the solution is evident from Equation (1.20); it is an oscillation with angular frequency ω_l but with amplitude that declines exponentially with time. Figure 1.4 shows the form of the solution for $C = 1$, $f/(2m) = 0.1\,\mathrm{s}^{-1}$, $\omega_l = 1\,\mathrm{s}^{-1}$ and $\phi = 0$.

The effect of light damping is to reduce the angular frequency ($\omega_l < \omega$) as well as to reduce the amplitude of the oscillation exponentially with time.

Exercise 1.4 A mass of $0.1\,\mathrm{kg}$ suspended on a spring is subjected to a force per unit displacement of $2\,\mathrm{N \cdot m^{-1}}$. There is light damping with $f = 0.1\,\mathrm{N \cdot m^{-1} \cdot s}$. What is the frequency of the oscillation of the mass and by what fraction will the original amplitude be reduced in 10 seconds?

1.2.2 *Heavy damping*

In this case we have $\frac{f^2}{4m^2} - \omega^2 > 0$ and Equation (1.17) is of the form

$$x = A \exp\left\{ -\left(\frac{f}{2m} - \sqrt{\frac{f^2}{4m^2} - \omega^2} \right) t \right\}$$

$$+ B \exp\left\{ -\left(\frac{f}{2m} + \sqrt{\frac{f^2}{4m^2} - \omega^2} \right) t \right\}. \qquad (1.21)$$

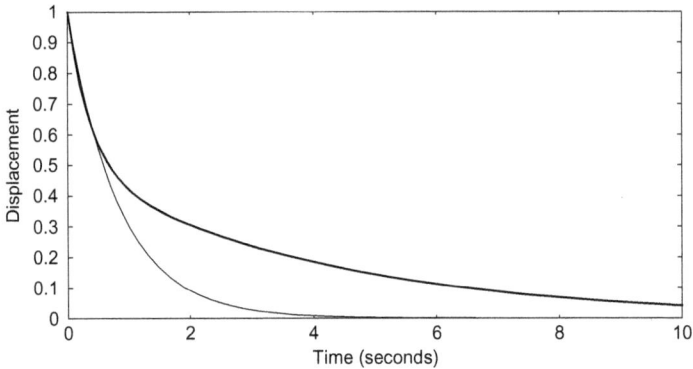

Figure 1.5 Examples of heavy damping (thick line) and critical damping (thin line).

Since $\frac{f}{2m} > \sqrt{\frac{f^2}{4m^2} - \omega^2}$, both terms in (1.21) are in the form of a declining exponential. The appearance of this function is shown in Figure 1.5, where now $f/(2m) = 1.5$ and $A = B = 0.5$, with other quantities as for Figure 1.4. From the figure it is seen that there is no oscillation. For the mass-on-a-spring example, if the whole equipment were immersed in treacle giving extremely high damping, the mass would smoothly move towards the equilibrium position and would not overshoot it.

1.2.3 *Critical damping*

Critical damping occurs when $\frac{f}{4m^2} - \omega^2 = \omega_l = 0$, which from Equation (1.20) with $\phi = 0$ gives

$$x = C \exp\left(-\frac{f}{2m}t\right). \qquad (1.22)$$

This function is shown in Figure 1.5 for $f/(2m) = 1$; it is the condition that gives the fastest smooth fall to the equilibrium position without overshooting it.

Having established the criterion of critical damping we can now change the nomenclature used in Sections 1.2.1 and 1.2.2 and refer to the more specific terms 'under-critical damping' rather than 'light damping' and 'over-critical damping' rather than 'heavy damping'.

Exercise 1.5 A mass experiences a force of 0.5 N m^{-1} when displaced from its equilibrium position. If it is critically damped and is displaced initially by 10 cm what will be the displacement after 5 seconds?

1.3 Forced Vibration and Resonance

We now consider a damped system that is being driven by an external periodic force of the form $F\cos(\omega_F t)$. The differential equation is now

$$\frac{d^2x}{dt^2} + \frac{f}{m}\frac{dx}{dt} + \omega^2 x = \frac{F}{m}\cos\omega_F t. \tag{1.23}$$

We seek a solution of the form

$$x = A\cos\omega_F t + B\sin\omega_F t. \tag{1.24a}$$

This gives

$$\frac{dx}{dt} = -A\omega_F\sin\omega_F t + B\omega_F\cos\omega_F t \tag{1.24b}$$

and

$$\frac{d^2x}{dt^2} = -A\omega_F^2\cos\omega_F t - B\omega_F^2\sin\omega_F t. \tag{1.24c}$$

We insert the expressions for x and its derivatives from (1.24) into Equation (1.23) and equate coefficients of $\cos(\omega_F t)$ and $\sin(\omega_F t)$ on the two sides of the equation, giving

$$m(\omega^2 - \omega_F^2)A + f\omega_F B = F \tag{1.25a}$$

and

$$-f\omega_F A + m(\omega^2 - \omega_F^2)B = 0. \tag{1.25b}$$

From (1.25b)

$$B = \frac{f\omega_F}{m(\omega^2 - \omega_F^2)}A. \tag{1.26}$$

Substituting this value of B in (1.25a) gives

$$A = \frac{m(\omega^2 - \omega_F^2)}{m^2(\omega^2 - \omega_F^2)^2 + f^2\omega_F^2}F \tag{1.27a}$$

and then

$$B = \frac{f\omega_F}{m^2(\omega^2 - \omega_F^2)^2 + f^2\omega_F^2} F. \tag{1.27b}$$

A solution in the form of (1.24a) can be converted into the form of Equation (1.3) by taking

$$C = \sqrt{A^2 + B^2} \tag{1.28a}$$

and

$$\phi = \tan^{-1}\left(\frac{-B}{A}\right) \tag{1.28b}$$

where, as previously indicated, the angle ambiguity of ϕ is resolved by noting that $\sin\phi$ has the sign of $-B$ and $\cos\phi$ has the sign of A. From this the solution of the differential equation (1.23) is

$$x = \frac{F}{\sqrt{m^2(\omega^2 - \omega_F^2)^2 + f^2\omega_F^2}} \cos(\omega_F t + \phi) \tag{1.29a}$$

where

$$\tan\phi = \frac{-f\omega_F}{m(\omega^2 - \omega_F^2)}. \tag{1.29b}$$

However, (1.29a) is not a complete solution of Equation (1.23). The solution of Equation (1.14), which is similar to (1.23) but with zero on the right-hand side, can be added to the solution given by (1.29a), and this combination will still be a solution of Equation (1.23). Whatever the nature of the added component (corresponding to under-critical damping, critical damping or over-critical damping), it will contain an exponentially declining factor with time so that eventually that component disappears leaving the periodic *steady-state solution* as given by Equation (1.29a). The component that eventually disappears is called a *transient*, meaning that it has a limited duration. In mathematical terminology the solution of differential Equation (1.14) is a *complementary function*, Equation (1.29a) gives the *particular solution* and the sum of the two is the *general solution*.

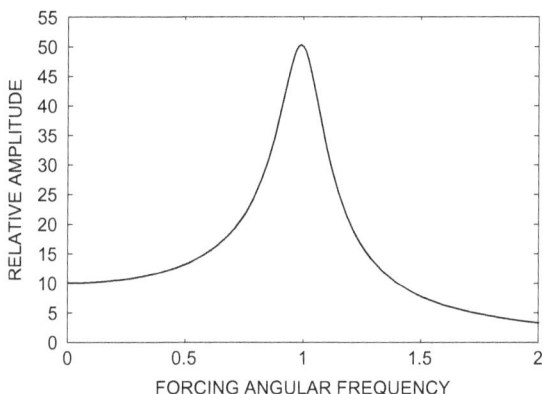

Figure 1.6 The resonance of a forced vibrator.

For a system with particular characteristics of mass m, restoring force per unit displacement κ, and resisting force per unit speed f, the amplitude of the steady-state solution depends on the applied angular frequency, ω_F. This dependence is shown in Figure 1.6 for $F = 10$, $m = 1$, $\omega = 1$ and $f = 0.2$. It is clear that there is a *resonance effect*, giving a maximum amplitude at an angular frequency close to $\omega_F = 1$, which is the angular frequency of the undamped and unforced simple harmonic oscillator.

To find the exact angular frequency for resonance we express the amplitude as

$$C = \frac{F}{\{m^2(\omega^2 - \omega_F^2)^2 + f^2\omega_F^2\}^{1/2}}. \tag{1.30}$$

For an angular frequency, ω_{res}, giving a resonant maximum amplitude we have $dC/d\omega_F = 0$ or

$$\left(\frac{dC}{d\omega_F}\right)_{\omega_F = \omega_{res}} = -\frac{\omega_{res}F\{f^2 - 2m^2(\omega^2 - \omega_{res}^2)\}}{\{m^2(\omega^2 - \omega_{res}^2)^2 + f^2\omega_{res}^2\}^{3/2}} = 0. \tag{1.31}$$

There are solutions to Equation (1.31) that are not maxima (i.e. $\omega_{res} = 0$) because ω_{res} is a factor in the numerator. Also, $\omega_{res} = \infty$ because the power of ω_{res} is higher in the denominator of (1.31) than in the numerator. It is clear from Figure 1.6 that these two values correspond to minima. The maximum is found from the solution of

$$f^2 - 2m^2(\omega^2 - \omega_{res}^2) = 0, \tag{1.32}$$

which gives

$$\omega_{res} = \sqrt{\omega^2 - \frac{f^2}{2m^2}}. \tag{1.33}$$

For the parameters that gave Figure 1.6 we find $\omega_{res} = 0.990\,\mathrm{s}^{-1}$, very close to the value of ω. For quite heavy damping, although still under-critical, with $f = 0.5$, $\omega_{res} = 0.935\,\mathrm{s}^{-1}$.

From Equation (1.30) we can see that for an undamped oscillator $(f = 0)$ with $\omega_F = \omega$ the amplitude would be infinite. Of course such a situation could not occur in practice — from Equation (1.12) it would correspond to the system having infinite energy — so what would happen is that the amplitude of the oscillation would steadily increase with time until the system broke down in some way.

We have now dealt with the essential mathematics to deal with a wide variety of situations, some occurring in everyday life and others only in a scientific context.

Exercise 1.6 A mass of 0.1 kg is subjected to a restoring force of $10\,\mathrm{N\ m}^{-1}$ when displaced and a damping force of $1\,\mathrm{N\ m}^{-1}$ s. It is subjected to an external force, $F\cos(5t)$, where t is in seconds. What is the steady-state amplitude and phase of the resultant vibration?

Problems 1

1.1 The figure shows a cylinder containing a piston of mass m, the weight of which is supported by the pressure, P, of the gas below it. The whole arrangement is contained within a vacuum enclosure. The piston is slightly displaced from its equilibrium position. Show that it will execute simple harmonic motion and find the frequency of that motion. The length of the cylinder below the piston is h and the temperature remains constant so the gas obeys Boyle's Law, $PV = \text{constant}$, where V is the volume of the gas. Ignore friction between the piston and cylinder.

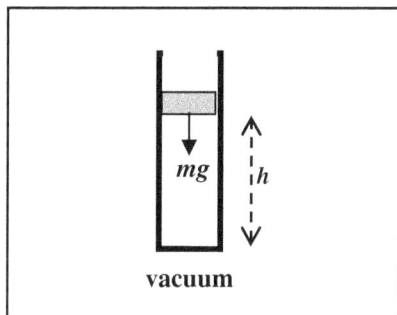

1.2

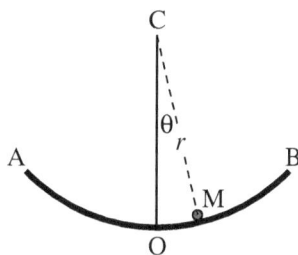

AB is part of a cylindrical surface of radius r with the point C on the cylinder axis. A small body M, of mass m, is slightly displaced from the point O and oscillates around that point. Due to friction between M and the surface of the cylinder it experiences a force opposing its direction of motion of magnitude fV, where f is a constant and V its speed. Determine the differential equation that describes its motion and the frequency with which it oscillates around the point O.

1.3

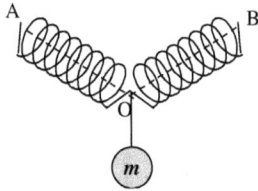

Two equal springs, each of length L and anchored at points A and B, support the mass m with the angle AOB equal to $120°$. The springs exert a force per unit change of length of κ. Show that if m is displaced downwards by a very small distance x ($x \ll L$) and then released it will undergo simple harmonic motion and determine the period of the oscillation.

[Note: For this problem you will need to apply the binomial theorem in the way described in Appendix 1.]

Chapter 2

Resonance in Everyday Life

2.1   A Girl on a Swing

A child's swing may be thought of as a recreational pendulum. Figure 2.1 shows a watercolour by the American artist Winslow Homer (1836–1910) where the girl is providing the motive power by bending her body to-and-fro; however, especially for younger children, it is customary for someone else, usually a parent, to provide the motive power by pushing at an appropriate time. The pusher stands at the back of the swing and every time the swing is momentarily stationary at the zenith of its backward motion a forward push is provided. Initially, when the swing has small amplitude of motion, the push will add energy to the swing and the amplitude will increase. Once the amplitude of the swing reaches a desired maximum level then gentle pushes are all that is required to maintain that amplitude and to compensate for the loss of energy during the previous cycle due to friction and air resistance.

What has been described is an everyday example of resonance. The driving force for the swing is not varying in sinusoidal fashion, as described in Equation (1.23), but is a sequence of impulsive forces, provided at times that will add energy to the motion. This is one example of resonance where the driving force is impulsive rather than sinusoidal, and this is quite common in natural occurrences of resonance phenomena.

Figure 2.1 A girl on a swing (Winslow Homer).

2.2   The Opera Singer and the Wine Glass

A typical wine glass is illustrated in Figure 2.2; for a good quality glass the bowl is not too thick, the stem is slender and it stands on a circular base.

If the glass is gently tapped, preferably with a hard object such a table knife, the glass will emit a pure musical sound that will quickly die away. A more sustained sound can be obtained by running a wet finger round the rim of the glass. As long as the finger motion is maintained the glass will emit a continuous pure tone. The glass is set into vibration with the rim of the glass executing a motion between the extreme positions marked by full and dotted lines in Figure 2.3. High-speed cine-cameras have been used to capture the motion and the distortion of the rim can be surprisingly large: that shown in Figure 2.3 is not an exaggeration.

The natural frequency of wine-glass vibrations — somewhere around middle C — is within the range of the human voice and if a singer, or some device for emitting sound, maintains that frequency in the vicinity of a wine glass then it will be set into vibration, although

Figure 2.2 A typical wine glass.

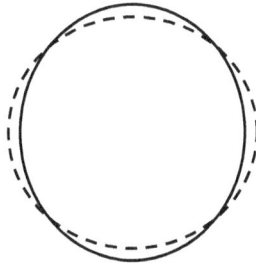

Figure 2.3 The limits of the vibration mode of the circular rim of a wine glass.

normally with a tiny amplitude. However, it has been reported that some past opera singers have been able to shatter wine glasses with the power of their voices; the Italian tenors Enrico Caruso (1873–1921) and Beniamino Gigli (1890–1957) have been mentioned in this respect, in stories that they were singing to entertain and that some wine glass in the vicinity shattered. This is quoted as an example of resonance where the forcing vibration is so strong that the strain in the glass due to the large amplitude of vibration was more than the glass could withstand. There are well-authenticated examples where a wine glass has been shattered by a human voice. If the wine glass is held close to the mouth, then someone with an exceptionally loud voice, who can maintain a steady note for long enough can do it. Actually it is easier to do with an expensive wine glass; high quality glass, with a higher refractive index that gives it a greater sparkle, tends to be less elastic and more brittle than the cheaper variety, and hence shatters more readily.

Although the opera singer versions of the breaking of wine glasses are doubtful (but perhaps just possible), the basis of the phenomenon is easily demonstrated, especially using sound-producing equipment that can maintain a fixed frequency at high intensity for long periods. In the 1970s there was an advertisement on an American television network that showed a wine glass being shattered by the eminent American jazz singer Ella Fitzgerald (1917–1996), but this was done by amplifying her voice using equipment that was a product of the company being advertised.

2.3   Bridges

The suspension bridge, constructed with strong cables held at each end by tall, sturdy towers to support the bridge deck, is a common type of bridge for spanning large distances; they have been built for many centuries to cross gorges in the mountainous regions of the Himalayas. A typical modern example in the UK is the Humber Bridge, completed in 1981, which spans the river Humber and links East Yorkshire to Lincolnshire (Figure 2.4), with a central span between the main suspension towers of 1,410 m in length. The world's longest central span for a suspension bridge, 1,991 m, is that of the Akashi–Kaikyo bridge in Japan, completed in 1998, which links the city of Kobe with Awaji island.

There are many designs of bridges — some built of brick, stone or concrete — that are rigid structures without much tendency to vibrate. Conversely, as will be appreciated from Figure 2.4, suspension bridges by their nature are very flexible and have a natural frequency of vibration; that being so they are prone to resonance effects.

Figure 2.4 The Humber Bridge.

2.3.1 *The Broughton Suspension Bridge*

On 14th April 1831, 74 soldiers of the 60th Rifle Corps of the British Army were returning from military exercises on Kersal Moor, a part of the city of Salford now integrated into Greater Manchester. Marching back to their barracks in Salford, four abreast, they started crossing the Broughton Suspension Bridge, completed in 1826, which crossed the River Irwell and linked the two Salford suburbs Pendleton and Broughton. As they were trained to do, they marched at the British Army regulation rate — 120 30$''$ (76 cm) paces per minute — and as they crossed the bridge it began to move in resonance with their marching rate. The sensation was not unpleasant so they broke into song and instinctively slightly adjusted their pace to more closely match the frequency of the rocking bridge. Then, just before they reached the end of the bridge, disaster struck. A link of a suspension chain snapped under the strain and one end of the bridge collapsed, hurling 40 men into the river. Fortunately, the river was not deep at the time and, although there were a few broken limbs and some minor injuries, there were no fatalities. Ever since then, whenever

British soldiers cross a bridge in marching formation they receive the order 'break step' and a smart column of men all marching in unison suddenly degenerates into a disorderly group all doing their own thing.

The lesson was learned and no more bridges have collapsed for that reason. The Albert Bridge over the River Thames in London, named after Queen Victoria's consort Prince Albert, was completed in 1872. At each end of the bridge there is a small notice, still in place, that reads 'All troops must break step when crossing this bridge'.

One of the greatest bridge disasters of all time was the collapse on 1st July 1940 of the Tacoma Narrows Bridge, a suspension bridge in Washington State, US, which spanned the Tacoma Narrows. The cause of the collapse was a high wind and is often described as being due to a resonance effect, but according to the engineers who analysed the disaster, this was not so; the effect was due to a phenomenon known as *aeroelastic flutter*, the force that causes a flag to flutter in a strong wind.

2.3.2 *The Millennium Bridge, London*

The London Millennium Bridge (Figure 2.5), completed in 2000, is a pedestrian suspension bridge crossing the River Thames, with the notable landmarks St Paul's Cathedral close to the northern end and the Tate Modern and Bankside art galleries and the reproduction Shakespearean Globe Theatre close to the southern end. When it opened on 10th June 2000, those crossing the bridge were aware of a somewhat unpleasant swaying motion. After two days of use the bridge was closed, and after two years of remedial work to correct the swaying motion it was reopened in 2002. Despite the correction of the initial fault it is still known, to Londoners, somewhat affectionately, as the 'Wobbly Bridge'.

Of course, the resonance effect that caused the swaying was not due to pedestrians all marching in step; it was due to a *positive feedback effect*, a reaction by the pedestrians to an original minor swaying motion. As the bridge swayed so those on the bridge swayed in unison, thus adding to the energy of the swaying motion and increasing

Figure 2.5 The London Millennium (Wobbly) Bridge showing St Paul's Cathedral at its northern end.

its amplitude. Since there were up to 2,000 people on the bridge at any one time, the effect of their combined reaction was significant. There was never any danger of the bridge collapsing but it was an unanticipated fault that cost 5 million pounds to correct.

2.4 Washboard Roads

In remote rural areas, particularly in large countries such as Australia or the United States, minor roads are often unpaved but nevertheless subjected to constant passage by motor vehicles. Such roads often exhibit a corrugation structure, as seen in Figure 2.6a, and are usually called *washboard roads*. The term 'washboard' applied to these roads is because their surfaces resemble that of a device, used more often in the past than now, for washing clothes by hand (Figure 2.6b). The action of rubbing the soaped article up and down the washboard removed the ingrained dirt and grime.

There is some uncertainty about the mechanism for producing a washboard surface on a dirt road. One theory is that the original flat road would have had some imperfection in the form of a

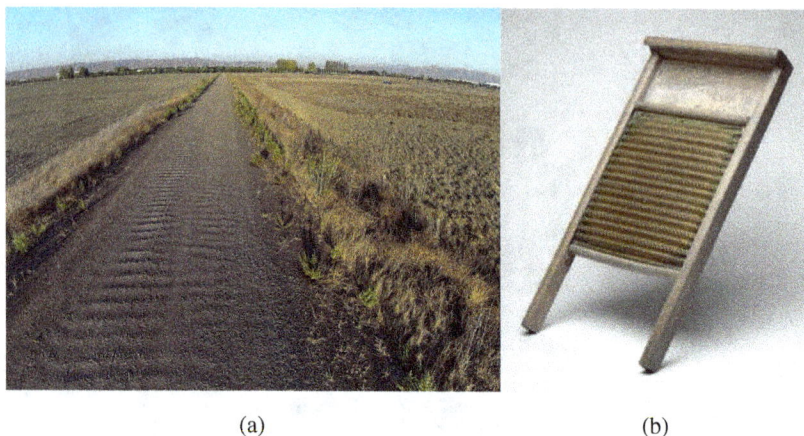

(a)   (b)

Figure 2.6 (a) A washboard road, Fremont, California. (b) A washboard.

slight depression. A motor vehicle passing over this depression would descend into it and slightly increase the depression, much as dropping an object into a soft dry surface will cause a depression, and the springiness of the tyres would then cause the vehicle to bounce upwards. The next time it descended it would once again make a depression and this process, continuously repeated, would have created a succession of depressions that deepened as more vehicles passed over the road. This mechanism — if indeed it is the explanation — would require that the springiness of tyres was similar in most vehicles and that they were all moving at similar speeds. What *has* been observed is that the distance between peaks, usually in the range 0.75–1.0 m, is larger for roads where average speeds are higher.

The washboard effect is also present on some railway lines, corresponding to very tiny undulations on the steel rails. Trains passing over such surfaces generate a loud noise known as *roaring rails* and similar effects have occurred with the overhead wires of electric trains and trams.

A vehicle driving on a corrugated road at a speed such that the natural frequency associated with the springiness of its tyres matches that at which it meets the depressions will vibrate rather unpleasantly, even to the extent of damaging the vehicle, as well as reinforcing the existing corrugations. This resonance effect can be lessened by either slowing down or speeding up to destroy the resonance; another effective way of reducing the effect is to lower the tyre pressure, which changes the natural frequency of the tyre bounce.

2.5 Buildings and Earthquakes

There are many regions in the world where major earthquakes may occur; when they occur it is without warning and, if in an occupied region, they may cause major structural damage and loss of life. Figure 2.7 shows a view of part of the city of San Francisco, in California, after an earthquake struck it in 1906, causing 3,000 deaths and destroying 80% of the city's buildings. It can be seen from the figure that the damage is severe, although the strongest earthquakes recorded have released more than 200 times more energy than that one.

Figure 2.7 A view of part of San Francisco after the 1906 earthquake.

A detailed examination of Figure 2.7 shows something quite interesting: many of the tallest structures seem little affected while smaller buildings are heavily damaged, if not completely demolished; this runs counter to one's instinct that tall buildings should be more vulnerable. Earthquakes vary a great deal in their characteristics in terms of the way that the ground shakes. Tremors can occur over short or long periods of time — up to almost a minute in severe events — and with shaking frequencies that can vary over a considerable range. Buildings themselves have a natural frequency of vibration that depends on their physical characteristics — for example, their height and breadth, as well as the type of construction. In the case of a squat building the frequency is very high and it would require a special investigation to determine. For example, if a small explosive charge were set off close by, then detectors attached to the building could measure the frequency of the small-amplitude vibrations that were induced in the building. For very tall buildings, the natural frequency is lower and the fact that they are flexible can be readily experienced. As an extreme example, the top of the tallest building in the world, the Burj Khalifa in Dubai with a height of 880 m, can sway by up to 1.7 m in a high wind and with this flexibility there is associated a natural frequency.

In 1985 there was an earthquake off the Pacific coast of Mexico that affected a part of Mexico City at a distance of 350 km. The energy generated by the earthquake was about twice that of the San Francisco earthquake, with at least 10,000 people killed and 50,000 injured (estimates vary) and many more being made homeless. It was found that very short and very tall buildings remained standing and were little damaged, while buildings between 6 and 15 stories high were the most vulnerable, either collapsing completely or sustaining considerable damage, because their natural frequencies were in the range of the frequencies being generated by the earthquake.

Efforts are made to protect buildings in earthquake-prone areas by installing large dampers, designed to reduce the amplitude of the vibrations, in critical locations — essentially to introduce a large value of f in Equation (1.30). Another approach is to follow the old Chinese maxim 'A wise tree bows before the wind' and to design the structure to accommodate large vibrational amplitudes without unduly compromising the complete structure.

2.6 Resonance and Musical Instruments

The full range of types of musical instrument is involved in the playing of music by a symphony orchestra. There are the *string instruments*, such as the violin, cello and harp, where the strings are set vibrating by stroking or plucking. Next we have the *woodwinds*, such as the flute and the clarinet, in which vibrations are set up in a column of air either by blowing across an open hole in the instrument, or by a blowing action that sets a reed into vibration. Related to woodwinds, in that vibrating columns of air produce the musical note, are *brass instruments*, which are typically, but not necessarily, made of brass. Here it is the vibration of the player's lips that play an important part of producing the right note. Finally we have *percussion instruments*, such as drums and triangles, which produce sounds when struck.

The performance of all these instruments owes much to their detailed design but here we shall just concentrate on the principles underlying two general types — those that depend on vibrating columns of air and those that depend on plucked strings — without delving into specific design characteristics.

2.6.1 *Resonance of air in a pipe*

An instrument that is not part of a typical symphony orchestra is the large and non-mobile *pipe organ*, which consists of banks of vertical pipes of different lengths. The instrument is played by an organist, who uses hands and feet to manipulate both a keyboard and a set of switches, called *stops*, which control the timbre and loudness of the notes being played. The movement of air in the pipes is the essential element in the action of a pipe organ and a source of high-pressure air is required, provided in older instruments by someone operating a large bellows, but these days by an electrically driven pump.

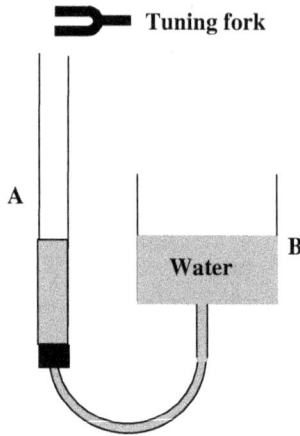

Figure 2.8 Producing resonance in a pipe.

An experiment that demostrates the action of an organ pipe is illustrated in Figure 2.8. If a tuning fork were set vibrating, then in general one would hear just the rather low intensity note coming from it. However, by raising or lowering the container B, the level of the water in tube A is changed and hence the length of the air column in the tube. If the air column in A is adjusted to the right level, then the natural frequency of air vibration in A matches that of the tuning fork and a loud sound is heard coming from A.

The way that the air in A is vibrating is shown in Figure 2.9. At the closed end of the pipe the air is stationary and not vibrating at all, a so-called *node* of the vibration, while at the open end the amplitude of vibration is at a maximum, an *antinode* of the vibration. Although the representation in Figure 2.9 makes it seem to be a *transverse wave*, where the motion of the material is perpendicular to the propagation direction, the actual vibration of the air molecules is <u>along</u> the length of the pipe; it is a *longitudinal wave*, where the motion is to-and-fro along the direction of propagation of the wave motion. The vibration in the pipe corresponds to one quarter of a complete wave so the wavelength of the vibration is

$$\lambda = 4l. \tag{2.1}$$

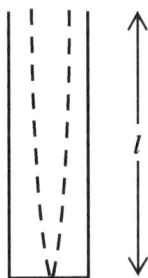

Figure 2.9 The pattern of air vibration in a pipe closed at one end.

For an acoustic wave the frequency, n, is connected to the wavelength by

$$n\lambda = c, \tag{2.2}$$

where c is the speed of sound in the medium in which the sound travels ($300\,\mathrm{m\cdot s^{-1}}$ for air). Hence, for the note middle C ($256\,\mathrm{Hz}$), the length of the pipe is

$$l = \lambda/4 = \frac{c}{4n} = \frac{300}{4 \times 256} = 0.293\,\mathrm{m}. \tag{2.3}$$

Another kind of resonance effect in a pipe occurs when the pipe is open at both ends, giving an antinode at each end and a pattern of vibration as shown in Figure 2.10. In this case the vibration in the pipe corresponds to one half of a wavelength and the length of a pipe that would resonate at middle C is $0.586\,\mathrm{m}$.

Figure 2.10 The pattern of air vibration in a pipe open at both ends.

In a pipe organ there are pipes of both the open and closed varieties, each with a different mode of activation: *flue pipes*, and *reed pipes*. The flue pipe is activated by blowing air over an opening at the end of the pipe which has a sharp edge called a *labium*. This produces a low-pressure region below the opening, just as a low-pressure region is produced above the wing of an aircraft by the flow of air over its upper surface. When the pressure in the pipe just under the opening falls sufficiently, the air-flow changes direction and moves under the labium. Eventually the pressure falls over the opening to once again divert the air-flow over the opening. This air-flow, alternating on either side of the labium, induces alternating high and low pressure waves within the pipe that set up the vibration of air that produces the note.

The reed in a reed pipe is a fine curved strip of springy brass that is situated at one end of the pipe. The shape of the reed is such that when air flows past the reed, low pressure is created below it that pulls the reed down against a hard surface. Now the air cannot flow past the lower surface of the reed so the low pressure is not maintained and the reed springs back to its original unstressed state. The beating frequency of the reed is matched to the natural frequency of the pipe and the air in the pipe is set into motion, giving a loud note.

The timbre, or quality, of the note depends on lower intensity frequencies that accompany the main frequency, which is also known as the *fundamental* or the *first harmonic*. These accompanying higher *harmonics* are notes at some, usually small, multiple of the main frequency. In the experiment shown in Figure 2.8, the pipe will also resonate to a frequency that is three times the tuning-fork frequency for the fundamental. Now there is three quarters of a complete wave within the pipe (with a node at the closed end, another node within the pipe and an antinode at the open end), and what is being generated is a *third harmonic* resonance effect. For the open pipe there can be an antinode at each end with a complete wavelength in the pipe so that in this case there can be a harmonic with twice the fundamental, or a *second harmonic*. Organ pipes, set into motion by either the flue or reed mechanisms, will emit not only the fundamental but also some harmonics that will affect the quality of the note being heard. The timbre is also affected by the nature of the material from which the pipe is constructed and by its shape; pipes with varying cross sections and of different widths generate different amounts of the harmonic components.

Many musical instruments operate on the same general principles as those described for the pipe organ. For example, the flute operates like a flue pipe: air is blown onto a sharp edge and the flute can be of either the closed- or open-pipe variety. The clarinet is a reed instrument that operates very similarly to a reed pipe in a pipe organ.

Exercise 2.1 The longest pipe in the Town Hall organ in Sydney, Australia, which is also the longest in the world, has a length of 19.5 m. If it is open at each end, what is the frequency of the sound it emits?

2.6.2 *Resonance in a string*

A string, which can be made of almost any material, under tension and firmly anchored at each end, will emit a note when plucked. If the plucking is done at the centre of the string then it will vibrate in the manner shown in Figure 2.11a. The frequency of the note emitted will be

$$n = \frac{1}{2l}\sqrt{\frac{T}{m}}, \tag{2.4}$$

where l is the length of the string, T the tension within it and m its mass per unit length.

By plucking it at different points, the string can also be set into vibration as in Figure 2.11b, which gives a frequency $2n$, the second harmonic, or as in Figure 2.11c, which gives a frequency $3n$, the third harmonic. In general, plucking the string, or drawing a bow across it, will set it into a form of vibration where it will most strongly emit the fundamental but with an admixture of higher harmonics that gives the rich interesting tone of a stringed musical instrument.

(a)

(b)

(c)

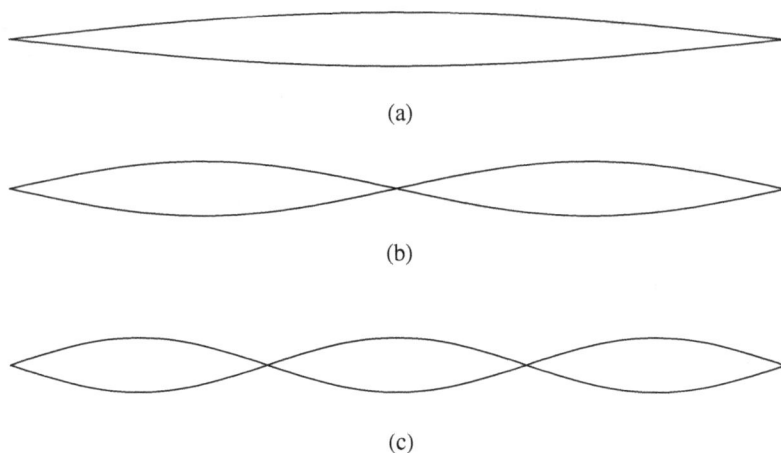

Figure 2.11 The vibration of a string. (a) Fundamental (b) Second harmonic (c) Third harmonic.

Exercise 2.2 A string of length 0.2 m with mass $1\,\mathrm{g\cdot cm^{-1}}$ is under a tension of 10 N. If it is plucked then what is the fundamental frequency and the frequencies of the next two harmonics it emits?

2.6.3 *The violin*

If a vibrating string were set up in the open air without being attached to an instrument, then the intensity of the sound emitted would be very low. The function of the main body of the musical instrument is to act as a resonator and to amplify the sound so that a wider audience can hear it. A violin, the most common stringed instrument in a symphony orchestra, can be played over a wide range of frequencies, from G3 (196 Hz) to E7 (2637 Hz) and resonance must be achieved over the complete range.

The body of the violin (Figure 2.12) contains two types of resonator: the wooden body and the air contained within it. At any particular frequency the total amplification will have contributions from both sources, although the net effect may be different from the sum of the two individual effects. The way that the two resonances combine will depend on their relative phases (i.e. the relationship

Figure 2.12 The front and side view of a violin.

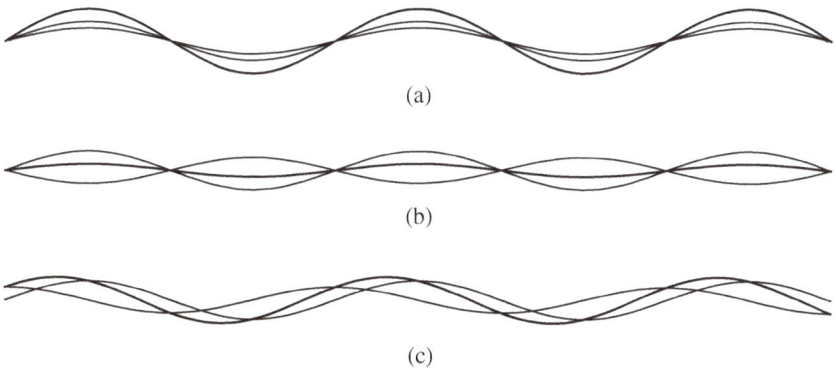

(a)

(b)

(c)

Figure 2.13 Combining two resonance outputs (thin lines) to give resultant (thick line) with phase differences (a) 0, (b) π and (c) $\pi/2$.

between their peaks and troughs). In Figure 2.13a the peaks and troughs of waves of different amplitudes coincide (phase difference zero) and the resultant wave has large amplitude. If the peaks of one output overlap the troughs of the other (phase difference π),

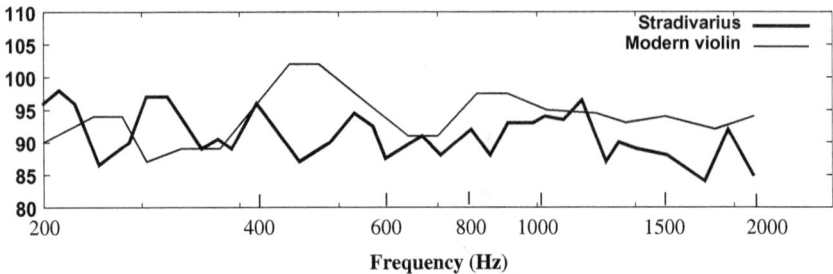

Figure 2.14 A comparison of the response at various frequencies for a modern violin with a Stradivarius.

then they will oppose each other and give a small resultant as shown in Figure 2.13b and an intermediate situation (phase difference $\pi/2$ in this case) is illustrated in Figure 2.13c.

For a complicated shape like a violin there will be several resonant frequencies, corresponding to different modes of vibration of both the wooden body and the contained air, and this makes the amplification more uniform over the whole frequency range. The intensity of the sound for a single frequency as a function of frequency is shown in Figure 2.14 both for a high quality modern violin and for a violin constructed by the eminent violin maker Antonio Stradivari (1644–1737). The Stradivarius violin gives a slightly more uniform response over the frequency range, especially in the lower part of the range up to 1,000 Hz — a small difference but clearly important for the quality of sound it gives.

There are many other aspects of violin construction that contribute to its quality but the resonance response of the violin is by far the most important.

Chapter 3

Electrical Circuits and Resonance

3.1 Direct-Current Circuits

For an electrical circuit the relationship between applied potential difference (PD), V, current, I, and resistance, R, is given by Ohm's Law:

$$V = IR. \tag{3.1}$$

Resistance is one type of *impedance*: the higher its value the more it impedes the motion of the electrons that constitute the current. With potential difference expressed in volts (V) and current in amperes (A) the unit of resistance is the ohm (Ω). The reciprocal of resistance is referred to as *conductance*, the unit of which is the *siemens* (S), sometimes called the *mho* (ohm in reverse). Thus a conductance of $(10\,\Omega)^{-1}$ is $100\,\text{mS}$.

Resistors can be linked in various ways. For two resistors with resistances R_1 and R_2 linked in series, as shown in Figure 3.1a, the combined resistance, R_C, is the sum of the two resistances or

$$R_C = R_1 + R_2. \tag{3.2}$$

If the resistors are linked in parallel, as in Figure 3.1b, then the combined resistance is given by

$$\frac{1}{R_C} = \frac{1}{R_1} + \frac{1}{R_2} \tag{3.3}$$

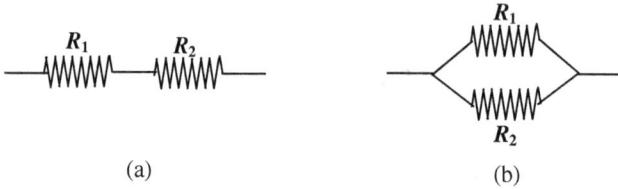

Figure 3.1 Linked resistors (a) in series and (b) in parallel.

or, in words, the combined conductance is the sum of the two conductances.

The resistor is an impedance for both an applied direct or alternating PD; now we shall consider impedances that only operate with an alternating PD.

Exercise 3.1 What is the combined resistance for two resistors of resistance 2Ω and 3Ω (a) in series and (b) in parallel?

3.2 Expressing an Alternating Potential Difference

The simplest way to express the form of an alternating PD is

$$V = V_0 \sin(\omega t), \tag{3.4}$$

where V_0 is the PD amplitude, t is time and ω is the *angular frequency* in radians·s^{-1}, linked to the frequency n (in Hz) as given by Equation (1.4).

Analysis involving alternating current circuits is made much simpler by the use of complex numbers, giving a PD expressed as

$$V = V_0 \exp(i\omega t) \tag{3.5a}$$

and, correspondingly, for an alternating current,

$$I = I_0 \exp(i\omega t) \tag{3.5b}$$

where

$$\exp(i\theta) = \cos(\theta) + i\sin(\theta). \tag{3.5c}$$

The modulus is the amplitude of the quantity (PD or current) and the argument of 'exp' is the phase of the quantity compared with that at $t = 0$.

The effect of advancing the phase by δ changes (3.4) to

$$V_0 \sin(\omega t + \delta) = V_0 \sin(\omega t) \cos(\delta) + V_0 \cos(\omega t) \sin(\delta) \qquad (3.6a)$$

and (3.5a) to

$$V_0 \exp\{i(\omega t + \delta)\} = V_0 \exp(i\omega t) \exp(i\delta). \qquad (3.6b)$$

The change for the complex expression is much simpler, just involving a product with a phase factor, $\exp(i\delta)$, and it is this property that gives the complex notation the advantage.

Exercise 3.2 Show that a phase shift of π converts a PD, V, into $-V$.

3.3 Complex Impedances

When the German physicist and mathematician Georg Ohm (1789–1854) first introduced Ohm's Law in 1827, it was only applied to real quantities, including resistance. Now we consider other kinds of impedance, which can be expressed in complex form, that come into play with alternating currents and for which Ohm's Law is still applicable.

3.3.1 *Inductors*

Inductors are impedance devices that depend on two physical phenomena. The first is that an electric current produces a magnetic field in its vicinity and, in particular, a current flowing through a coil gives a magnetic field, directed along the axis of the coil, of strength proportional to the current (Figure 3.2). The second is that a varying magnetic field passing through a coil will induce a potential difference across it of magnitude proportional to the rate of change of magnetic field.

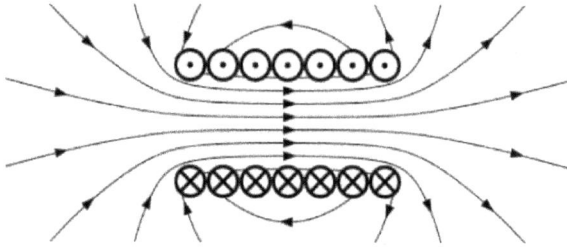

Figure 3.2 Magnetic flux through a coil carrying a current. The dots indicate current flowing out of the page and crosses current flowing into the page.

We now imagine that an alternating current passes through a coil of zero resistance, so that there is no potential difference across the coil due to resistance. The alternating current produces an alternating — and hence varying — magnetic field within the coil that, in its turn, induces an alternating potential difference across it. The self-induced potential difference will, at any instant, be proportional to the rate of change of the current. The constant of proportionality depends on the nature of the coil and is known as the *self-inductance*, L. This gives the potential difference across the coil as

$$V = L\frac{dI}{dt},$$ (3.7a)

or

$$V_0 \exp i\omega t = L\frac{dI}{dt}.$$ (3.7b)

The solution of the differential Equation (3.7b) is

$$I = \frac{V_0}{i\omega L}\exp(i\omega t).$$ (3.8)

For comparison with Ohm's Law, Equation (3.1), we express (3.8) as

$$V = i\omega L I$$ (3.9a)

from which it is seen that the inductor is equivalent to a *complex impedance* $z = i\omega L$ and

$$V = zI.$$ (3.9b)

This is similar to Ohm's Law with the complex PD, current and impedance replacing the real quantities in the direct PD application.

In addition, the rules for combining complex impedances in series and in parallel are precisely those given in Equations (3.2) and (3.3).

If we replace i by the expression

$$i = \cos\frac{\pi}{2} + i\sin\frac{\pi}{2} = \exp\left(i\frac{\pi}{2}\right), \qquad (3.10a)$$

then, since $i^2 = -1$, we have $1/i = -i$ and

$$\frac{1}{i} = \cos\frac{\pi}{2} - i\sin\frac{\pi}{2} = \exp\left(-i\frac{\pi}{2}\right). \qquad (3.10b)$$

From (3.8) and (3.10b)

$$I = \frac{V_0}{\omega L}\exp i\left(\omega t - \frac{\pi}{2}\right). \qquad (3.11)$$

We see that a PD of form (3.5a) applied across an inductor of self-inductance L gives an alternating current of amplitude $V_0/(\omega L)$ that lags in phase $\pi/2$ behind the applied PD. The SI unit of inductance is the *henry* (H).

Another form of inductance is when an alternating current in one coil induces an alternating PD across another coil to which it is magnetically linked. The linkage between the two coils is described by their *mutual inductance, M*.

Exercise 3.3 What is the total impedance if two inductors of inductance 2 H and 5 H are (a) in series and (b) in parallel if the applied PD has an angular frequency of 100 radians·s^{-1}? (Note: your answer will be in ohms.)

3.3.2 *Capacitors*

In its most basic form a *capacitor* consists of two conductors separated by a dielectric material (Figure 3.3), A dielectric is an insulator that, under the influence of a PD applied across it, undergoes a charge displacement (i.e. a shift in its electron content in the direction of the field but without any current flow). This induces equal and opposite charges on the conductors, of amount Q on one conductor and $-Q$ on the other. The electric field at any point between the conductors will be proportional to Q, as will the potential difference,

Dielectric

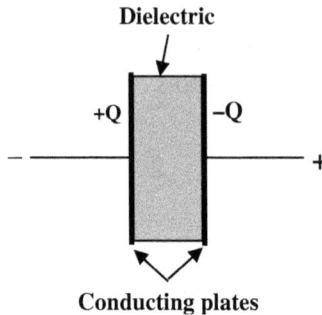

Conducting plates

Figure 3.3 A schematic capacitor. The signs on the wires indicate the direction of the applied PD.

V, between the conductors. The capacitance of the capacitor, C, is defined as the charge carried by the conductors per unit potential difference between them, or $C = Q/V$, which gives

$$V = \frac{Q}{C}. \tag{3.12}$$

An expression can be found linking capacitance and potential difference in terms of current rather than charge by differentiating both sides of (3.12) with respect to time, giving

$$\frac{dV}{dt} = \frac{1}{C}\frac{dQ}{dt}, \tag{3.13a}$$

Since the change of charge δQ in an interval δt corresponds to the flow of a current in that interval we have $\delta Q = I\delta t$ which, from Equation (3.13a), gives

$$\frac{dV}{dt} = \frac{I}{C}. \tag{3.13b}$$

From (3.5a) and (3.13b)

$$i\omega V_0 \exp i\omega t = \frac{I}{C}$$

or

$$V = \frac{1}{i\omega C}I = -\frac{i}{\omega C}I, \tag{3.14}$$

showing that the capacitor is equivalent to a complex impedance $-\frac{i}{\omega C}$.

We also find by using Equation (3.10a) and rearranging,

$$I = \omega C V_0 \exp i \left(\omega t + \frac{\pi}{2} \right), \tag{3.15}$$

showing that a PD of form (3.5a) applied across a capacitor C gives an alternating current of amplitude $\omega C V_0$ that leads the applied PD in phase by $\pi/2$. The SI unit of capacitance is the farad (F).

Exercise 3.4 What is the combined impedance of two capacitors of capacity 10^{-3} F and 3×10^{-3} F (a) in series and (b) in parallel if the applied PD has an angular frequency of 100 radians·s^{-1}? (Note: your answer will be in ohms.)

3.4 A Series LCR Resonance Circuit

We now consider a simple series circuit containing a resistor, an inductor and a capacitor driven by an alternating potential difference, as shown in Figure 3.4. Since the impedance elements are in series, the total impedance, Z, is found by adding the separate impedances, just as for simple resistors. This gives

$$Z = R + i \left(\omega L - \frac{1}{\omega C} \right). \tag{3.16}$$

By analogy with Ohm's Law we now write

$$V = IZ$$

or

$$I = \frac{V_0 \exp(i\omega t)}{R + i \left(\omega L - \frac{1}{\omega C} \right)}. \tag{3.17}$$

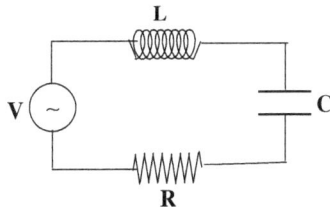

Figure 3.4 An LCR series circuit.

The impedance part of this expression is rationalized (i.e. put in the form $a + ib$) by multiplying the top and bottom of the right-hand side of (3.17) by the complex conjugate[1] of the divisor, which gives

$$I = \frac{V_0 \exp(i\omega t)\left\{R - i\left(\omega L - \frac{1}{\omega C}\right)\right\}}{R^2 + \left(\omega L - \frac{1}{\omega C}\right)^2}. \tag{3.18}$$

We make the transformation

$$\frac{R}{\sqrt{R^2 + \left(\omega L - \frac{1}{\omega C}\right)^2}} = \cos(\delta) \quad \text{and} \quad \frac{\omega L - \frac{1}{\omega C}}{\sqrt{R^2 + \left(\omega L - \frac{1}{\omega C}\right)^2}} = \sin(\delta). \tag{3.19}$$

Note that this transformation satisfies the condition $\cos^2 \delta + \sin^2 \delta = 1$ and defines δ by

$$\tan(\delta) = \frac{\omega L - \frac{1}{\omega C}}{R} \tag{3.20}$$

where $\sin(\delta)$ and $\cos(\delta)$ have the signs of the numerator and divisor of (3.20) respectively. In place of Equation (3.18) we can now write

$$I = \frac{V_0 \exp i(\omega t - \delta)}{\sqrt{R^2 + \left(\omega L - \frac{1}{\omega C}\right)^2}}, \tag{3.21}$$

from which we see that the impedances in series are equivalent to a single impedance of magnitude $\sqrt{R^2 + (\omega L - \frac{1}{\omega C})^2}$ that gives a phase shift of $-\delta$ where δ is defined by Equation (3.20).

For a PD of given amplitude the current will depend on its angular frequency and will clearly be a maximum for angular frequency ω_0 when

$$\omega_0 L - \frac{1}{\omega_0 C} = 0$$

or

$$\omega_0 = \frac{1}{\sqrt{LC}}. \tag{3.22}$$

---

[1]The complex conjugate of $a + ib$ is $a - ib$ and $(a + ib)(a - ib) = a^2 + b^2$.

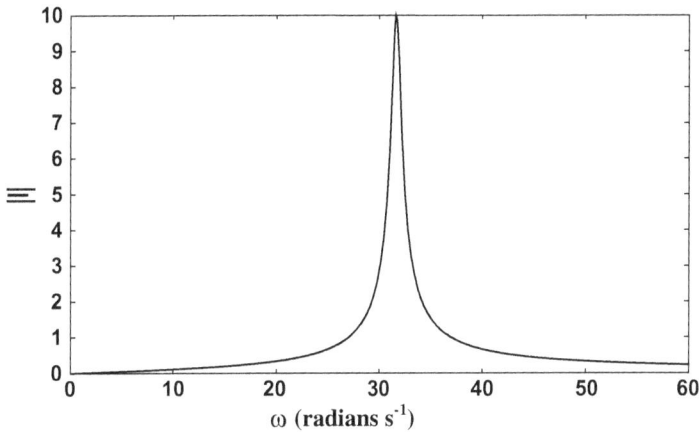

Figure 3.5 The magnitude of the current through a series LCR circuit as a function of the frequency of the applied PD.

This is a resonance condition when the applied frequency equals the natural frequency of the circuit. The relationship between the magnitude of I, $|I|$, and ω is shown in Figure 3.5 for a circuit with $V = 10\,\text{V}$, $R = 1\,\Omega$, $L = 1\,\text{H}$ and $C = 10^{-3}\,\text{F}$; the similarity to the resonance curve shown in Figure 1.6 is evident.

A circuit of this kind can be used as a *tuning device* for receiving radio signals or as a *filter* to pick out particular frequencies from a signal with a wide range of frequencies.

Exercise 3.5 A resistor of resistance $5\,\Omega$, a capacitor of capacitance $10^{-3}\,\text{F}$ and an inductor of inductance $0.2\,\text{H}$ are arranged in series. What is the combined impedance and phase shift for an applied PD of angular frequency 100π radians per second (frequency $50\,\text{Hz}$)?

3.5   A Parallel LCR Resonance Circuit

Figure 3.6 shows a circuit with an inductor and capacitor in parallel, with a resistor in the inductor arm since any inductor is bound to have some, however little, resistance.

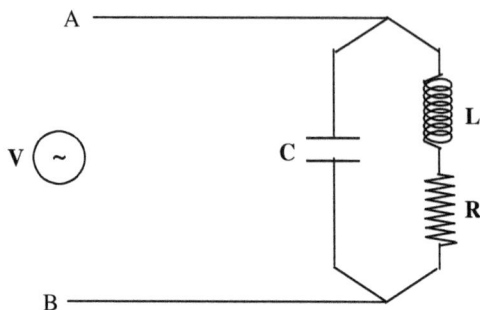

Figure 3.6 A parallel LCR circuit.

For two impedances z_1 and z_2, complex or otherwise, in parallel the overall impedance, Z, is given by

$$\frac{1}{Z} = \frac{1}{z_1} + \frac{1}{z_2}. \tag{3.23}$$

For the circuit shown in Figure 3.7 we have

$$\frac{1}{Z} = \frac{1}{R + i\omega L} - \frac{\omega C}{i} = \frac{1}{R + i\omega L} + i\omega C. \tag{3.24}$$

Rationalizing the first term on the right-hand side gives

$$\frac{1}{Z} = \frac{R - i\omega L}{R^2 + \omega^2 L^2} + i\omega C$$

$$= \frac{1}{R^2 + \omega^2 L^2}\{R + i\omega[C(R^2 + \omega^2 L^2) - L]\}. \tag{3.25}$$

This parallel LCR circuit has quite different properties from the series circuit so here we consider how the impedance of the circuit varies with frequency, viz.:

$$Z = \frac{R^2 + \omega^2 L^2}{R + i\omega\{C(R^2 + \omega^2 L^2) - L\}}. \tag{3.26}$$

To find the magnitude of the impedance we use the relationship that

$$\text{if } Z = \frac{z_1}{z_2} \quad \text{then} \quad |Z| = \frac{|z_1|}{|z_2|}$$

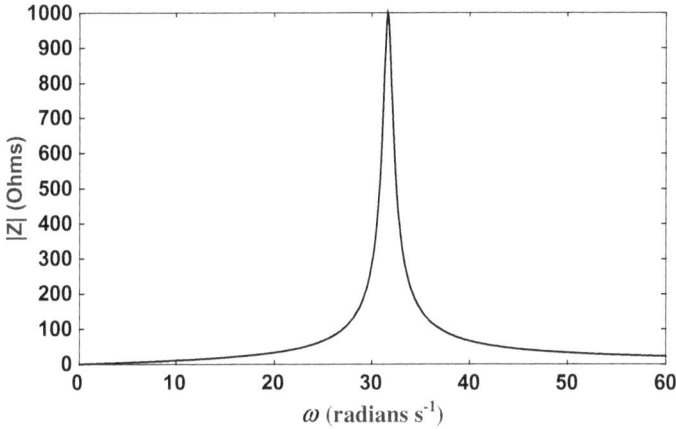

Figure 3.7 The variation of impedance with angular frequency for a parallel LCR circuit.

so that

$$|Z| = \frac{R^2 + \omega^2 L^2}{\{R^2 + \omega^2[C(R^2 + \omega^2 L^2) - L]^2\}^{1/2}}. \tag{3.27}$$

A useful transformation, using (3.22), is to write

$$\frac{1}{LC} = \omega_0^2, \tag{3.28}$$

where ω_0 is the resonant angular frequency for the series LCR circuit. This gives

$$|Z| = \frac{R^2 + \omega^2 L^2}{\{R^2 + \omega^2 C^2[R^2 + L^2(\omega^2 - \omega_0^2)]^2\}^{1/2}}. \tag{3.29}$$

For $R = 0$ and $\omega = \omega_0$ we have $|Z| = \infty$.

Figure 3.7 shows the variation of the magnitude of the impedance with angular frequency using the values that gave Figure 3.5, $R = 1\,\Omega$, $L = 1$ and $C = 10^{-3}\,\text{F}$.

If we consider the ideal case of $R = 0$ and $\omega = \omega_0$, so that $|Z| = \infty$, it means that there would be no current in the main circuit, (i.e. between points A and B in Figure 3.6). However that does not mean that there are no currents in the parallel arms of the circuit.

The PD between points A and B is given by Equation (3.5a), the current flowing through the inductor arm by Equation (3.11) and that through the capacitor arm by Equation (3.15). Replacing ω by ω_0 in (3.11) and using Equation (3.28) we find

$$I_L = \frac{V_0}{\omega_0 L} \exp i \left(\omega t - \frac{\pi}{2} \right) = \omega_0 C V_0 \exp i \left(\omega t - \frac{\pi}{2} \right)$$

$$= -\omega_0 C V_0 \exp \left(\omega t + \frac{\pi}{2} \right) = -I_C, \tag{3.30}$$

where I_L and I_C are the currents through the inductor and capacitor respectively. There are alternating currents in both arms of the circuit but they are equal in magnitude and differ in phase by π so they combine to give zero net current in the main circuit. If the resistance is not zero then this balance, in both magnitude and phase, will not be precise so the cancellation will be partial and the impedance will be finite, as shown in Figure 3.7, giving a net current through the main circuit.

A parallel LCR circuit can be used in many ways, for example as a *band-stop filter*. For many electrical circuits involved in sound reproduction the frequency of the mains supply can produce a hum that corrupts the acoustic signal. For a power supply at 50 Hz, as in the UK, the hum will be at 100 Hz (there are two intensity peaks for each cycle). A band-stop filter with high impedance for the range 49–51 Hz will largely remove the hum. There may be some degradation of the audio signal, although frequencies at around 100 Hz are not very important for most musical instruments and removing a small frequency range around 100 Hz would not be noticeable for speech, which is normally at much higher frequencies.

Exercise 3.6 An inductor of inductance 0.2 H and resistance 1 Ω is in parallel with a capacitor of capacitance 2×10^{-3} F. What is the maximum magnitude of the impedance of the arrangement, assuming that it occurs for angular frequency $(LC)^{-1/2}$?

Problems 3

3.1

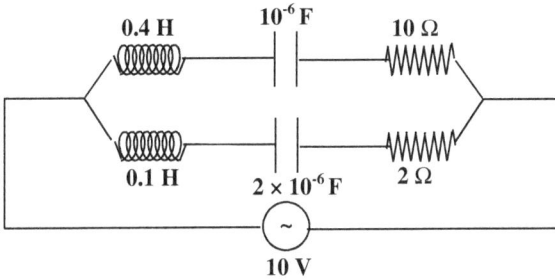

Find the current in each of the parallel arms of the circuit shown, in terms of the angular frequency of the applied PD. By considering separately the real and imaginary parts of these currents, excluding the time-dependent factor, $\exp(i\omega t)$, find the magnitude of the current in the main circuit, $|I_T|$.

Write a computer program to find the resonance conditions in this circuit by finding $|I_T|$ as a function of the frequency of the applied PD. You need only consider the range 10–500 Hz.

3.2

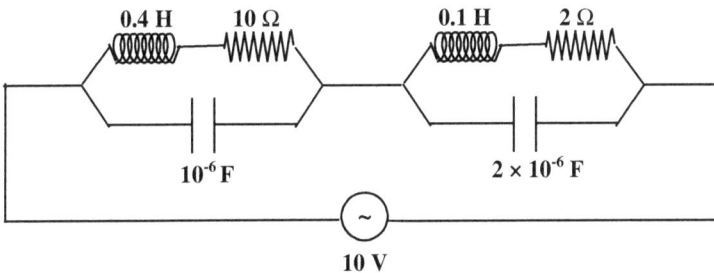

Find the real and imaginary components of the impedance of the circuit shown in Figure 3.6 in terms of the angular frequency of the applied PD. Hence find an expression for the amplitude of the impedance, $|Z_T|$, of the two units in series shown in the figure.

Write a computer program to find the amplitude of the total impedance, $|Z_T|$, as a function of the frequency of the applied PD. You need only consider the range 10–500 Hz.

Chapter 4

Resonance in the Solar System

A characteristic of all bodies in the Solar System, with the exception of the Sun, is that they are in orbit about some other body. Each orbit has an associated period and the periods of some pairs or triplets of bodies bear simple relationships to one another. These are examples of resonance, but not of a kind we have previously met.

4.1 Kirkwood Gaps

The planets known up to the late eighteenth century were those that could be visually observed, those out to Saturn. In 1772 the German astronomer Johann Bode (1747–1826) noticed that the planetary mean orbital radii out to Saturn closely followed a simple mathematical progression, which became know as Bode's Law. The orbital radii of planets are usually expressed in *astronomical units* (au), the mean distance of the Earth from the Sun (1.496×10^8 km), and, in units of au, Bode's Law could be expressed as

$$r_n = 0.4 + 0.3 \times 2^n \tag{4.1}$$

where the leading term, 0.4 au, is close to the radius of Mercury's orbit and for the remaining planets $n = 0, 1 \ldots$ for Venus, Earth, etc. The fit between Bode's Law and the actual mean orbital radii is shown in Table 4.1 and it will be noted that there is an apparent gap in the planets between Mars and Jupiter, corresponding to $n = 3$. When William Herschel (1738–1822) discovered Uranus in 1781 and

Table 4.1 Illustration of Bode's Law.

Planet (n)	Mean orbital radius (au)	Bode's Law
Mercury	0.387	0.4
Venus (0)	0.723	0.7
Earth (1)	1.000	1.0
Mars (2)	1.524	1.6
(3)		2.8
Jupiter (4)	5.203	5.2
Saturn (5)	9.539	10.0
Uranus (6)	19.19	19.6

its orbital radius was found to fit the law, it was generally believed that the law had some fundamental validity, although without any physical basis being established for it.

In 1846 the planet Neptune was discovered and its mean orbital radius, 30.07 au, did not fit the Bode's Law value, 38.8 au, and thereafter Bode's Law was much less highly regarded. However, in the late eighteenth century the gap for $n = 3$ in the Bode's Law progression was taken to indicate that there was a missing planet and astronomers began searching for it. On 1st January 1801 a small body was found in the right location, with a mean orbital radius of 2.77 au by the Italian astronomer, Giussepe Piazzi (1746–1826), the Director of the Palermo Observatory, who called it Ceres, after the guardian god of his native Sicily. Although somewhat small compared to other planets, with a diameter of 1,000 km, it satisfactorily filled the gap in Bode's Law. The neatness of the apparent completion of the family of planets was disturbed by the discovery of other similar, if somewhat smaller, bodies over the next few years at similar distances from the Sun and this was the prelude of many other discoveries of small bodies that are now known as *asteroids*, the search for which continues to the present day. They have a variety of sizes, from a few tens of metres to several hundreds of kilometres in mean diameter, and many tens of thousands have been discovered. They mostly occupy the zone between Mars and Jupiter, referred to as the *asteroid belt*,

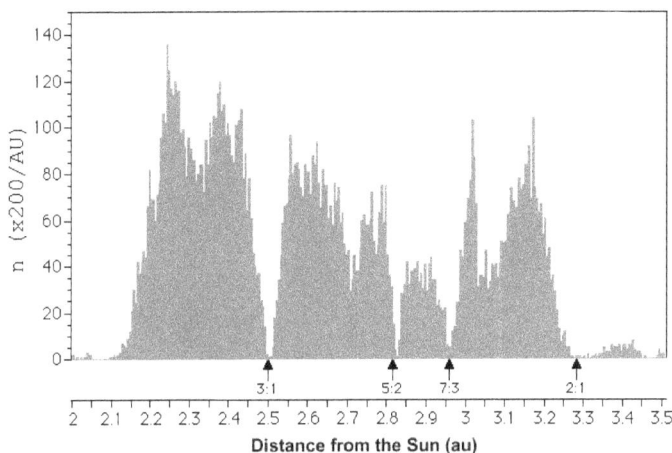

Figure 4.1 The distribution of asteroid semi-major axes showing Kirkwood gaps.

but there are some with orbits that take them within the orbit of the Earth and others that have been observed out as far as the region between Saturn and Uranus.

A diagram giving the distribution of asteroid mean orbital radii, such as Figure 4.1, indicates that there are many prominent gaps.

The American astronomer Daniel Kirkwood (1814–1895) explained these gaps in 1866. He noted that there are two very prominent gaps corresponding to one-third and one-half of Jupiter's period, which suggested that these gaps were a manifestation of some resonance phenomenon. An asteroid with one-half the period of Jupiter will make two complete orbits while Jupiter is making one. Thus the two bodies will always be closest together in the same region of the asteroid's orbit, so that the perturbation by Jupiter at closest approach will always modify the asteroid's orbit in the same way. The period of the asteroid will change increasingly in one direction until the asteroid's and Jupiter's orbits are sufficiently out of resonance for the nearest approaches, and hence maximum perturbations, to occur throughout the asteroid's orbit, rather than in one region, which greatly diminishes Jupiter's effect. For the one-third resonance the asteroid is perturbed at two points on opposite sides

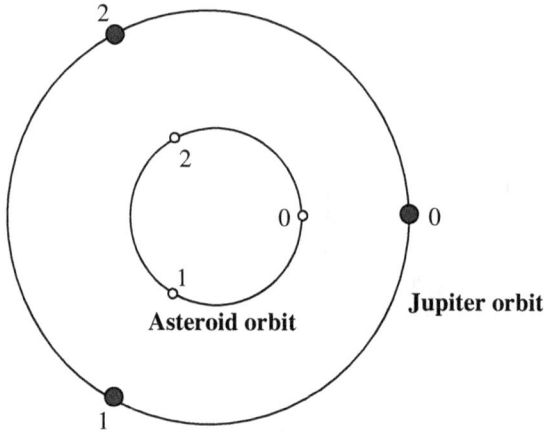

Figure 4.2 Consecutive closest approaches of asteroid and Jupiter for period ratio 2:5.

of its orbit. As increasing numbers of asteroid orbits were determined other *Kirkwood gaps* became more clearly seen and gaps at two-fifths and three-sevenths of Jupiter's period are also shown in the figure.

These gaps have evolved by the accumulation of tiny perturbations of the orbits over the age of the Solar System, almost 4.5 billion years, but it is possible to simulate their formation with a simple computer program; a FORTRAN program that performs this calculation is listed in Appendix II.

To see the pattern of how closest approaches occur, we consider in Figure 4.2 a case where the asteroid orbit has two-fifths of the orbital period of Jupiter.

The point marked 0 is a closest approach and the next one is at the points marked 1, where Jupiter has gone through $\frac{2}{3}$ of an orbit and the asteroid has gone through $1\frac{2}{3}$ — a ratio of 2:5. The next closest approach is at the points 2 where, from the points 1, once again Jupiter has gone through $\frac{2}{3}$ of an orbit and the asteroid through $1\frac{2}{3}$ of its orbit. The next closest approach is back at point 0 and the pattern of closest approaches is repeated. There have been three close approaches at different points in a span of two Jupiter periods. This can be generalized to when the ratio of the orbital period of the asteroid to that of Jupiter is m/n, where m and n are integers. In m orbital periods of Jupiter there will be $n - m$ close approaches at points uniformly distributed around the asteroid's orbit.

The factors that govern the distinctness of a Kirkwood gap are:

- The total elapsed time, i.e. 4.5 billion years for the Solar System.
- The number of closest-approach points, i.e. $n - m$; the effect is stronger for fewer closest approach points.
- The closeness of the asteroid orbit to that of Jupiter.

In one simulation using the program in Appendix II, 2,000 asteroids, taken with zero mass, were placed in circular orbits with radii uniformly distributed between 2.0 au and 4.5 au. Jupiter was placed in a circular orbit of radius 5.2 au and its position could be found analytically at any simulated time. The differential equations were solved for the motions of the asteroids under the gravitational influence of the Sun and Jupiter for a period of 10,000 years. Because of the short simulation time the full extent of the gaps is not reproduced, although where the perturbation effects are strongest they appear quite distinctly. The distribution of actual asteroid orbital radii tends to be concentrated between 2 au and 3.5 au, as indicated in Figure 4.1, but the simulation includes many asteroids closer to Jupiter. The results of the simulation are shown in Figure 4.3. The gaps are especially distinct for asteroids with orbits close to that of Jupiter but less distinct at larger distances.

When the periods of two bodies are in a simple ratio of two, usually small, integers they are said to be *commensurate*. Interactions between commensurate orbits are common in astronomy and we now describe other examples.

Exercise 4.1 Integers are associated with the planets from Mercury out to Neptune such that $n = 0$ for Mercury and $n = 8$ for Neptune with no planet associated with $n = 4$. Plot $\log(r_n)$ against n, where r_n is the orbital radius in au. Draw the best straight line through the points and from the line deduce a relationship of the form $r_n = ab^n$. Compare the values of r_n deduced from the line with the true values.

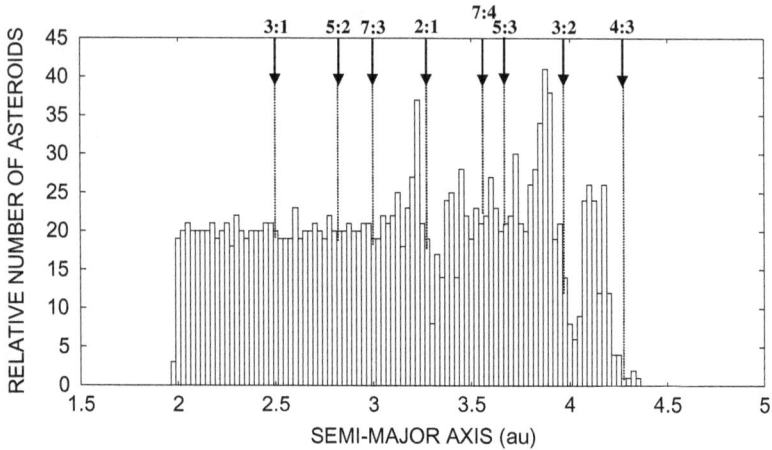

Figure 4.3 Simulating perturbation by Jupiter of asteroids with orbital radii originally uniformly distributed between 2.0 au and 4.5 au. The ratios are that of the period of Jupiter to that of an asteroid.

Exercise 4.2 In Figure 4.1 there are two depressions in the distribution at distances of 2.71 au and 3.04 au from the Sun. The period P is related to orbital radius r for a solar orbit by $P = 2\pi\{r^3/(GM)\}^{1/2}$ where G is the gravitational constant and M the mass of the Sun. Find the periods corresponding to these depressions and find the best commensurate relationships with the period of Jupiter, 11.862 years. You will find it convenient to use the solar-system units, described in Appendix II.

4.2 Saturn's Rings

In 1610, while using his newly acquired telescope to observe the planet Saturn, Galileo Galilei saw appendages that seemed like a pair of ears. What he was observing were Saturn's rings but the quality of his telescope was too poor to show a clear image. Galileo was astonished when in 1612 the appendages disappeared; Saturn had moved position so that the thin rings were edge-on to his line of sight, but he did not know that.

Modern telescopes show that Saturn's rings are quite detailed in structure (Figure 4.4). The dominant features are two bright rings,

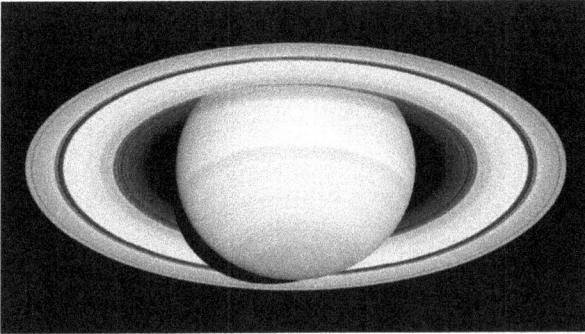

Figure 4.4 A high-quality telescope image of Saturn's rings (NASA/ESA/Hubble Space Telescope).

an outer ring A and an inner ring B, separated by a prominent dark band, the Cassini Division. Closer to Saturn there is a fainter ring C, difficult to see and referred to as the Crepe Ring. When Clerk Maxwell (1831–1879) was a student at Cambridge in 1857, he showed in an Adams Prize Essay that, for stability, the rings had to consist of small solid bodies independently orbiting Saturn. Infrared and radar data has shown that they are either icy bodies or perhaps ice-covered silicate bodies, varying in size from being small grains up to having diameters of several metres.

The Cassini division is produced by the same kind of perturbation that gives the Kirkwood gaps. In this system Saturn is the central body; the bodies forming the ring system correspond to asteroids; and some larger satellites of Saturn play the role of Jupiter in producing the Cassini Division and other less prominent gaps in the rings. Saturn has many satellites; the larger ones that are close enough to Saturn to appreciably perturb the ring bodies, with their diameters in km and orbital radii in units of 10^3 km, are shown in Table 4.2.

The position of the Cassini Division, which is very broad, corresponds to one half the period of Mimas, one third that of Enceladus, a quarter that of Tethys and one sixth that of Dione. Detailed observation has shown other fine divisions, which turn out to be commensurate with the periods of either Mimas or Enceladus (Figure 4.5). It is

Table 4.2 The diameters and orbital radii of
some larger Saturn satellites.

Satellite	Diameter (km)	Orbital Radii (10^3 km)
Mimas	397	185
Enceladus	504	238
Tethys	1,066	295
Dione	1,123	378

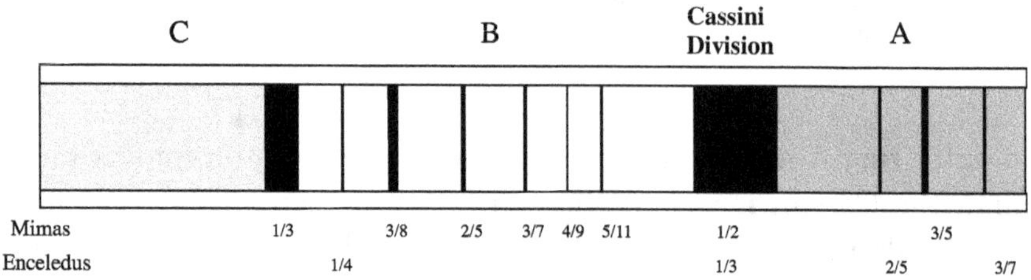

Figure 4.5 A representation of the main divisions in Saturn's rings showing their commensurabilities with the periods of Mimas and Enceladus.

clear from the relationship of several satellites to the Cassini Division that there should be commensurabilities between the orbital periods of those satellites. This is discussed further in Section 4.4.3.

The best images of Saturn's rings have been obtained by spacecraft. Figure 4.6 shows a false-colour picture taken by Voyager I, the false colours being used to give better contrast and to enable fine detail to be seen more easily. The ring system is found to consist of hundreds of separate rings and gaps of varying thickness. Outside the outer boundary of ring A there are three more faint rings, F, G and E that are not seen from Earth. Within the C ring there is a very faint D ring with a rather indeterminate inner edge. Although the main features of the ring system, as shown in Figure 4.5, are well understood, the precise nature of the mechanics that maintains the fine structure, as seen in Figure 4.6, is not understood, although some factors that operate are known. For example, the stability of the F ring is due to the presence of two small satellites, Pandora and

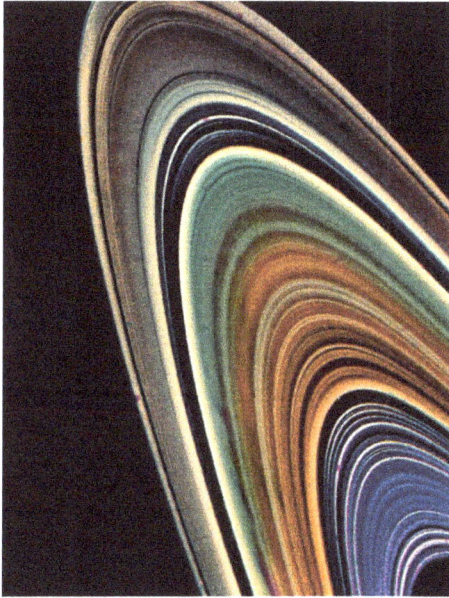

Figure 4.6 A false-colour view of Saturn's rings (NASA).

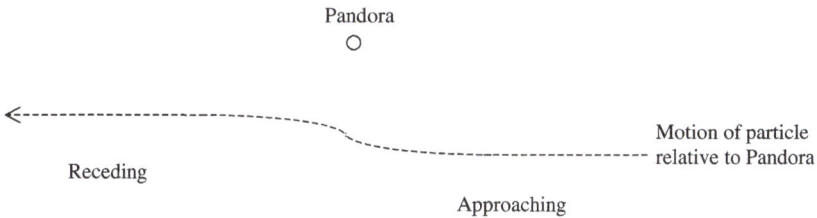

Figure 4.7 The motion of a ring particle relative to Pandora and within its orbit. On approach the particle gains energy and on receding loses energy. The loss slightly exceeds the gain because it is a little closer.

Prometheus that straddle the F ring and prevent it from spreading either inwards or outwards.

The action of these *shepherd satellites* can be explained fairly simply. Figure 4.7 shows the motion of a particle in an orbit inside that of Pandora, the outer of the two shepherds. The particle has a higher angular speed than Pandora and when it is behind the particle, but catching up, it will gain energy from the satellite while at the same time being deflected slightly towards Pandora. After the particle passes the satellite it is pulled back and loses energy and to a first approximation the gain and loss of energy cancel each other. However, since in the energy-losing part of the motion the particle is slightly nearer Pandora, the energy lost is marginally larger than that gained. The loss of energy corresponds to an orbit further inwards so the effect of Pandora is to repel the particle away from itself. By a similar argument it can be shown that if the particle's orbit is outside that of Prometheus and it is orbiting more slowly, then it is repelled by Prometheus, thus constraining it to have an orbit further outward. The net effect of the shepherds is that particles orbiting between them feel a pinch effect that prevents them from diffusing either inwards or outwards.

4.3   Volcanoes on the Satellite Io

When Galileo acquired his telescope in 1610, one of the first bodies he looked at was Jupiter. He discovered four satellites, the first to be observed in the Solar System other than the Moon. It is now known, from subsequent telescope and spacecraft observations, that Jupiter has 63 satellites. The ones seen by Galileo are the largest and are called the *Galilean satellites*; in order outwards from Jupiter, they are Io, Europa, Ganymede and Callisto. Io and Europa are, respectively, slightly more and less massive than the Moon and Ganymede and Callisto are, respectively, slightly larger and smaller than the planet Mercury although, because of their low densities, they are both considerably less massive than that planet.

Just before the Voyager 1 spacecraft reached Io in 1979, a paper by S.J. Peale, P. Cassen and R.T. Reynolds appeared in the journal *Science* predicting that there would be volcanoes on the satellite. This bold prediction was verified and Figure 4.8 shows the plume from the volcano, Pele. Many other volcanoes were subsequently discovered on Io.

The basis of the prediction was that the orbital periods of the neighbouring satellites Io and Europa are commensurate with periods in the ratio 1:2. Because of this, Europa and Io are always

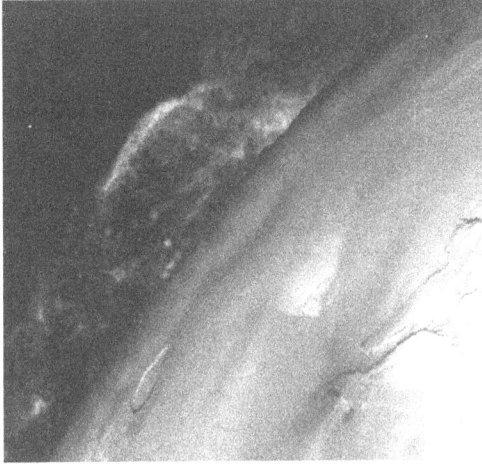

Figure 4.8 Plume from the Io volcano, Pele (NASA).

closest at the same point of Io's orbit. The perturbations to Io's orbit by Europa are thus enhanced by a resonance effect and the orbit becomes slightly eccentric. This eccentricity is very small but given Jupiter's large mass and its proximity to Io it is enough to give a periodic tidal flexing (i.e. stretching and contraction) of Io. We now describe the physics that produces heat from the tidal flexing.

4.3.1 *Elastic hysteresis and Q values*

For a perfectly elastic body following *Hooke's Law*, the amount by which it is stretched, known as the *displacement*, is proportional to the stretching force (Figure 4.9a). The total work done in stretching by an amount $\varepsilon_{\max}$ is $1/2\, F_{\max}\, \varepsilon_{\max}$, where $F_{\max}$ is the maximum force, and this is the shaded area under the line OP. If the force is then reduced back to zero the successive states of the rod are again represented by points on the line OP.

Most materials are not perfectly elastic and when being stretched and then contracting follow the paths indicated in Figure 4.9b. The work done *on* the material when being stretched is the area under the curve OAP and the work done *by* the material in returning to the unstressed state is the area under the curve OBP. The difference

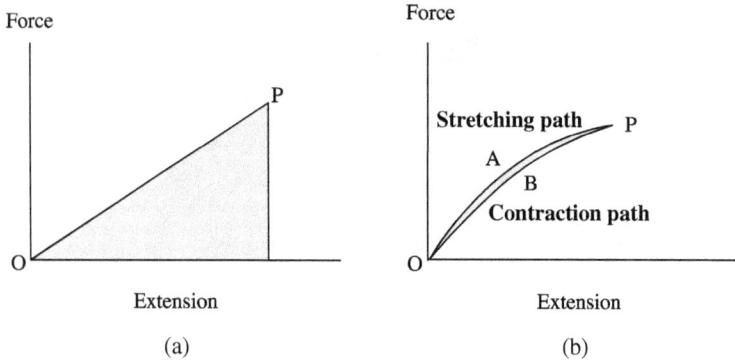

Figure 4.9 (a) The relationship between force and extension for a perfectly elastic material. (b) Extension followed by contraction for an imperfectly elastic material. The hysteresis heating corresponds to the shaded area.

is represented by the shaded area and appears as heat within the material. Internal friction between different grains within the material has generated heat during stretching and, from the principle of conservation of energy, the material cannot do as much work in contraction as was done to it in expansion. This phenomenon is called *hysteresis* and something similar occurs when magnetic materials are subjected to an alternating magnetizing field – *magnetic hysteresis*.

The extent of the hysteresis is described by the *quality factor, Q,* which can be defined in our present case as

$$Q = 2\pi \text{ (Energy stored in stretched body/Energy lost as heat)}.$$

Highly compressed planetary material might have Q values as large as 1,000, indicating that less than 1% of the maximum energy stored when it goes through a single stress cycle appears as heat. However, material in a lesser state of compression, close to the surface of a large body or within a small body, could have a Q value of 100 so that about 6% of the stored energy appears as heat.

4.3.2 *Elliptical orbits*

Here we briefly describe the characteristics of an ellipse (Figure 4.10), which is the form of the orbit of a planet or satellite about its

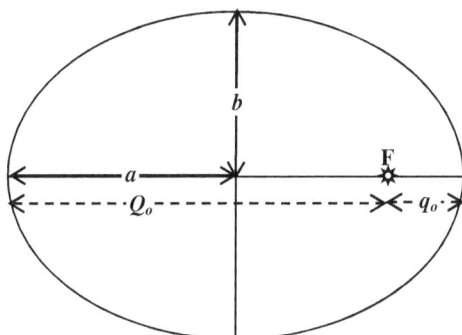

Figure 4.10 The geometry of an ellipse.

primary body. The point F is the *focus* occupied by the primary body (the Sun or a planet), a is the *semi-major axis* and b the *semi-minor axis* of the ellipse. The equation of an ellipse in Cartesian coordinates is

$$\frac{x^2}{a^2} + \frac{y^2}{b^2} = 1. \tag{4.2}$$

The quantities a and b are related through the *eccentricity*, e, of the ellipse by

$$b^2 = a^2 \left(1 - e^2\right). \tag{4.3}$$

For an ellipse with $0 \le e < 1$, $e = 0$ corresponds to a circle and as e increases the ellipse becomes more relatively stretched out along the major axis (i.e. the ratio b/a becomes smaller).

The closest approach of a planet to the Sun, q_0, is called the *perihelion distance* and is given by

$$q_0 = a(1 - e). \tag{4.4a}$$

The furthest distance, Q_0 (not to be confused with quality factor), is called the *aphelion distance* for an orbiting planet and is given by

$$Q_0 = a(1 + e). \tag{4.4b}$$

4.3.3 *The generation of energy in Io by tidal stressing*

We can make a ballpark estimate of the heating of Io by a rough calculation. A representation of Io, with radius r, is shown in Figure 4.11 divided into two equal hemispheres by a diametrical plane perpendicular to the orbital radius vector. The tidal force due to Jupiter stretches the satellite, with each hemisphere experiencing forces away from the centre of the satellite. To estimate the total stretching force we calculate the acceleration (force per unit mass) at P *relative to* O where P is a distance $1/2\,r$ from O. We then assume that this acceleration applies to all the mass in that hemisphere.

The acceleration at any point on the satellite consists of two parts: a gravitational component due to Jupiter and a centrifugal component due to its orbit of radius R around Jupiter. The accelerations at points O and P are

$$A_O = -\frac{GM_J}{R^2} + R\omega^2 \tag{4.5a}$$

and

$$A_P = -\frac{GM_J}{(R - r/2)^2} + (R - r/2)\omega^2, \tag{4.5b}$$

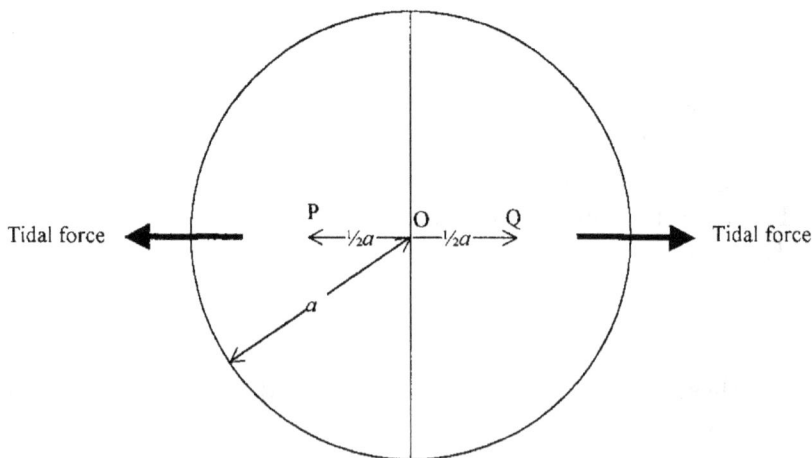

Figure 4.11 Simulation of the periodic tidal force acting on Io divided into two hemispheres.

where G is the gravitational constant, M_J the mass of Jupiter and ω the angular velocity of Io's orbit, which the same for all points of Io since it will always present the same face to Jupiter. Hence the acceleration of P relative to O is

$$A_{PO} = A_P - A_O = -\frac{GM_J}{(R - r/2)^2} + \frac{GM_J}{R^2} + (R - r/2)\,\omega^2 - R\omega^2. \quad (4.6)$$

Eliminating ω^2 by

$$\omega^2 = \frac{GM_J}{R^3}$$

and using the binomial theorem approximation with $a \ll R$ (Appendix I) yields

$$A_{PO} = -\frac{3GM_J r}{2R^3}. \quad (4.7)$$

Assuming that this is the average force per unit mass for the nearer hemisphere, the total force on it relative to O towards Jupiter is

$$F = \frac{3GM_I M_J r}{4R^3}, \quad (4.8)$$

where M_I is the mass of Io; the force on the far hemisphere is equal and opposite.

As the next approximation we approximate the satellite as a cube of side $2r$ being stretched along one principal direction. If Young's modulus for the satellite material is Y then the extension is

$$\varepsilon = \frac{F \times length}{Y \times area} = \frac{F}{2rY}$$

and the energy associated with the stretching is

$$E_S = \frac{1}{2} F\varepsilon = \left(\frac{3GM_I M_J}{4}\right)^2 \frac{r}{4Y} \times \frac{1}{R^6}. \quad (4.9)$$

The distance R varies in a cyclic fashion between $R_I(1 + e)$ and R_I $(1 - e)$ where R_I is the mean orbital radius of Io and e the eccentricity of its orbit. Hence the total change in energy between the extreme positions is

$$\Delta E_S = \left(\frac{3GM_JM_I}{4}\right)^2 \frac{r}{4Y} \times \frac{6}{R_I^7}\Delta R$$

$$= -\frac{27}{16}(GM_IM_j)^2 \frac{re}{R_I^6 Y}, \qquad (4.10)$$

since $\Delta R = 2R_Ie$.

From the definition of the quality factor the heat energy generated in each complete orbit of the satellite is $2\pi\Delta E_S/Q$, so that the energy dissipated within the satellite per unit time is given by

$$W = -\frac{2\pi\Delta E_S}{QP} = \frac{27}{8}\frac{(GM_IM_J)^2 \pi re}{R_I^6 QPY}, \qquad (4.11a)$$

where P is the orbital period. The reference to P can be removed since

$$P = 2\pi\sqrt{\frac{R^3}{GM_J}},$$

giving

$$W = \frac{27}{16}\frac{(GM_J)^{5/2} M_I^2 re}{R^{15/2}QY}. \qquad (4.11b)$$

The orbit of Io is very close to circular with $e = 0.004$. Minerals are very rigid materials with high values of Young's modulus, and when under high pressure they are even more rigid. For this reason we take a $Y = 10^{11} \mathrm{N\ m}^{-2}$. On the grounds that satellite material is not as highly compressed as planetary material but will be much more compressed than near-surface material on Earth, we take $Q = 500$. With these values we find $W = 2.6 \times 10^{14}$ W, which is a good ballpark estimate, given the usually accepted range of between 0.6 and 1.6×10^{14} W.

Exercise 4.3 If the radius of the orbit of Io were increased by 10% from its present value and its eccentricity increased to 0.02 then by what factor would the energy generation within it change due to tidal flexing?

4.4 Commensurabilities of Planetary and Satellite Orbits

The Kirkwood gaps and the gaps in Saturn's rings show that under some circumstances the relationship between the orbits of the lighter of two very disparate bodies (e.g. one a low-mass asteroid and the other a massive planet like Jupiter) is unstable if their periods are commensurate. Yet there are many other circumstances within the Solar System where commensurabilities, or very near commensurabilities, have been established between bodies of *comparable* mass.

4.4.1 *Planetary commensurabilities*

While Bode's Law is no longer regarded as significant in describing relationships between planetary orbits there are, notwithstanding, some striking commensurate relationships between the periods of major planets. Thus

$$\frac{\text{Period of Saturn}}{\text{Period of Jupiter}} = \frac{29.458 \text{ years}}{11.8623 \text{ years}} = 2.483 \approx \frac{5}{2}$$

$$\frac{\text{Period of Neptune}}{\text{Period of Uranus}} = \frac{164.79 \text{ years}}{84.01 \text{ years}} = 1.962 \approx \frac{2}{1}$$

In 1996, M.D. Melita and M.M. Woolfson described a mechanism by which these commensurabilities could become established during the period just after planets have formed, when they are moving in a resisting medium in the form of a disk around the Sun. Initially, if an orbit is eccentric and inclined to the mean plane of the disk, it will round-off (i.e. become more circular), decay (i.e. become smaller in extent) and reduce in inclination. We are considering the time when the orbits are close to circular and slowly spiralling in towards the Sun; this process continues until the resisting medium finally disperses, after which the orbits are more or less stable, apart from minor perturbations due to the gravitational effects of the planets on each other. The process by which commensurability is achieved is best illustrated by a computational approach.

For one of the computational models studied by Melita and Woolfson, two bodies were placed in orbit in a resisting medium, the inner one with the mass of Jupiter and the outer with the mass of Saturn. The initial eccentricities and inclinations of both orbits were 0.1 and 0.06 radians (3.4°) respectively. The initial ratio of the orbital periods was set at 2.5, close to 2.48, the present ratio between Saturn and Jupiter. Three runs of the computation were made with different initial relative positions in their respective orbits. Figure 4.12 shows the variation with time of the semi-major axis, eccentricity and inclination of the orbits of the two bodies. Both the semi-major axis and the inclination fall monotonically while the eccentricity falls at first but then rises again for both bodies. What is not readily seen in the figure is that the ratio of the two periods departs from 2.5 and settles down close to 2.0, actually oscillating about 2.02.

As with many numerical studies of complex systems, the results are easier to explain in terms of what is happening in the computation than to predict *ab initio*. When the computation is begun, the first effect is that the eccentricities fall quite quickly, as is seen in Figures 4.12c and d. A fall in eccentricity reduces the speed of a planet relative to the resisting medium (which is also in orbit around the Sun), which reduces the rate of energy dissipation and hence the rate of change of orbital parameters. The differential decay changes the ratio of the periods of the orbits until it becomes close to 2.0 — which means that, in a relative sense, Saturn's orbit has been decaying more rapidly than that of Jupiter. The relative rate of decay of the bodies will depend both on their masses and the distribution of the resisting medium.

At this stage a new process comes into play. We know that Kirkwood gaps in the asteroid belt correspond to periods commensurate with that of Jupiter and that the gaps in Saturn's rings correspond to simple commensurabilities with the periods of Mimas and Tethys. The reason for this is that the main orbiting body, Jupiter around the Sun or a close satellite around Saturn, *removes* energy from *interior* bodies in commensurate orbits. Conversely, although there is no obvious demonstration of this in the Solar System, energy is *added* to *exterior* bodies in commensurate orbits. Another effect

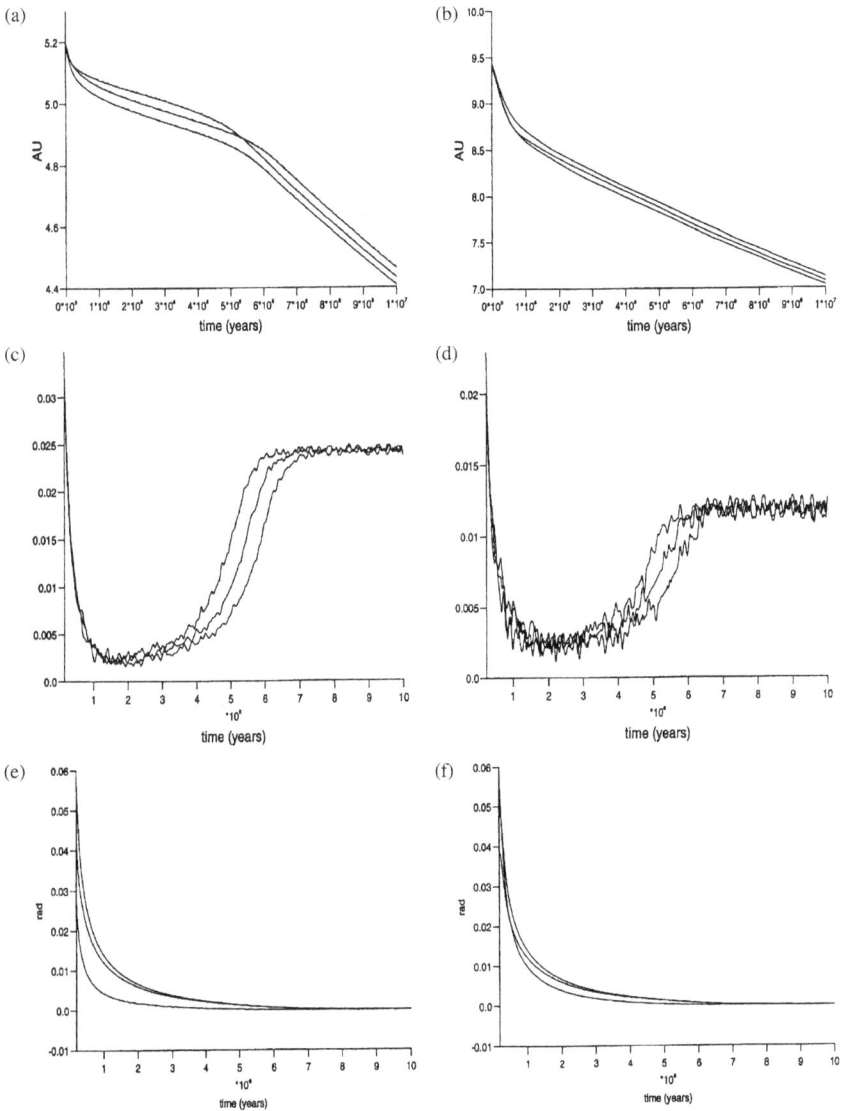

Figure 4.12 Semi-major axes, eccentricities and inclinations for Jupiter–Saturn systems starting near the 2:5 resonance in three different relative positions. (a) Semi-major axis: Jupiter, (b) Semi-major axis: Saturn, (c) Eccentricity: Jupiter, (d) Eccentricity: Saturn, (e) Inclination: Jupiter, (f) Inclination: Saturn.

of commensurability is that a resonance effect is set up such that the major mutual perturbations, corresponding to when the bodies are closest together, occur repeatedly at the same points of the orbits and amplify the effects of the perturbations. This increases the eccentricity which, in its turn, increases the rate of dissipation and hence the rate of decline in the semi-major axis. This latter effect is seen for Jupiter in Figure 4.12a by a change of slope of da/dt but it is not seen for Saturn. The reason for this is that the increasing dissipation due to increasing eccentricity of Saturn's orbit is balanced by a gain of energy due to the 2:1 resonance. Thus the rate of change of Saturn's semi-major axis is little affected while that of Jupiter is greatly increased. A non-obvious result of this is that the ratio of the periods, which had fallen from 5:2 to 2:1, now becomes locked at 2:1 although both orbits are still decaying (Figure 4.13).

The numerical experiments were repeated for the Jupiter–Saturn system with different initial eccentricities — the nine combinations with each eccentricity having the possible values 0.1, 0.2

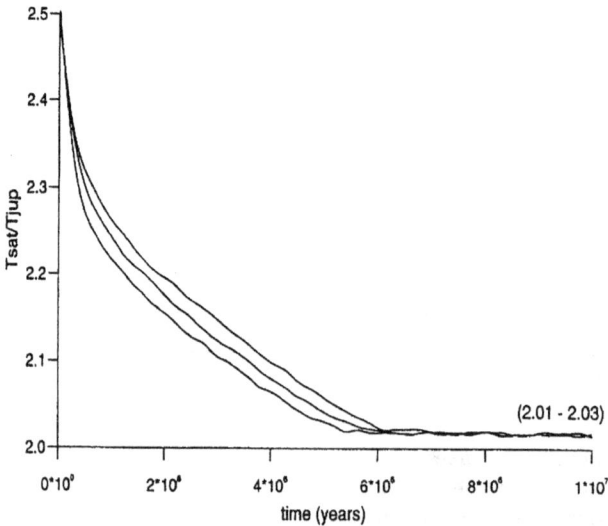

Figure 4.13 Ratio of periods for Saturn:Jupiter starting near the 5:2 resonance. Each curve represents a different initial relative position.

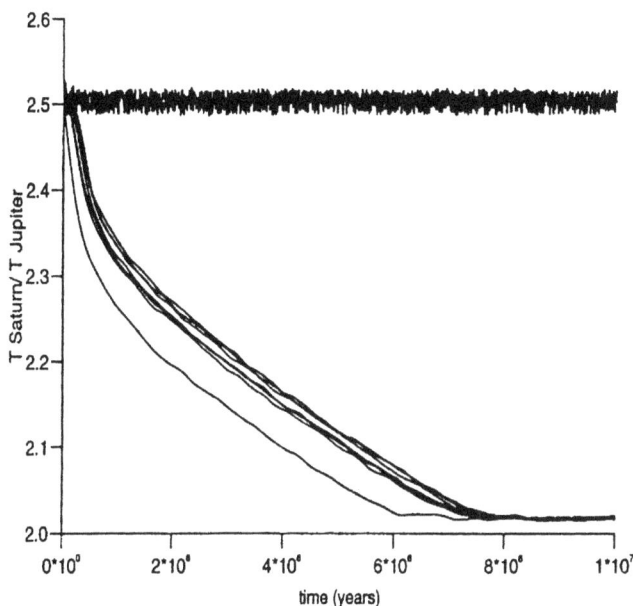

Figure 4.14 Evolution of the Jupiter–Saturn system for different pairs of initial eccentricities.

and 0.3 — but always with an initial orbital period ratio 2.5. The results are shown in Figure 4.14. Five of the combinations end up with ratio close to 2.0 and four stay at the original ratio 2.5.

Other trials with a Uranus–Neptune system gave a ratio of periods close to 2.0 starting with the present observed ratio of 1.96 (Figure 4.15). A factor that has not been taken into account in the calculations is the dissipation of the resisting medium that would take place in a few million years. As the medium is removed, so the rate of change of orbital parameters declines and will cease once the medium is removed completely. For this reason some pairs of orbital periods may become stranded away from the commensurabilities towards which they were evolving. Thus if the orbit of Uranus was decaying more slowly than that of Saturn then the ratio of the periods would gradually increase. The present ratio of the periods is 2.85, which could indicate that it was going towards 3.0 but was terminated by the removal of the medium.

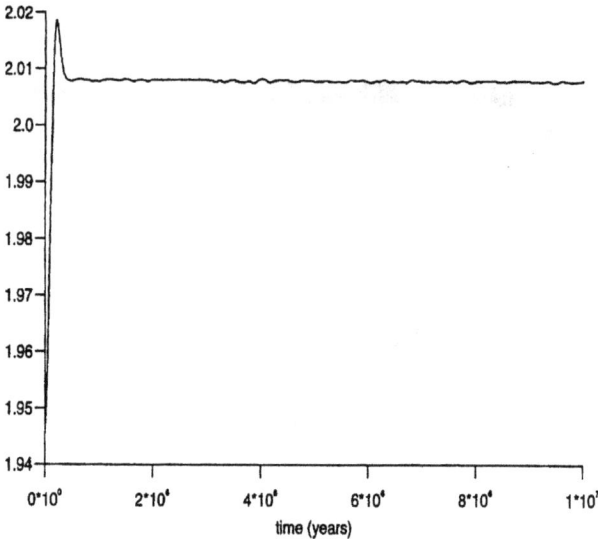

Figure 4.15 The Uranus–Neptune system starting near the present ratio of periods. The system reaches a resonant configuration with a ratio of periods slightly greater than 2.

Exercise 4.4 The radius of the orbits of Jupiter, Saturn and Uranus are, respectively, 5.20 au, 9.54 au and 19.19 au and their masses, in Earth units, are 317.8, 95.2 and 14.5. What is the ratio of the maximum gravitational force of Jupiter on Uranus to that of Saturn on Uranus? Find the ratio of the orbital period of Uranus to that of Jupiter. Compare the percentage departure from commensurability for the pair Uranus:Jupiter with that of the pairs Saturn:Jupiter and Neptune:Uranus.

4.4.2   *The commensurabilities of the Galilean satellites*

The periods of the orbits of Io, Europa and Ganymede are close to being in the ratio 1:2:4. Although these ratios are not precise what *is* precise is the relationship

$$n_1 - 3n_2 + 2n_3 = 0, \qquad (4.12)$$

where n indicates the *mean motion*, or average angular speed in the orbit, and subscripts 1, 2 and 3 correspond to Io, Europa and

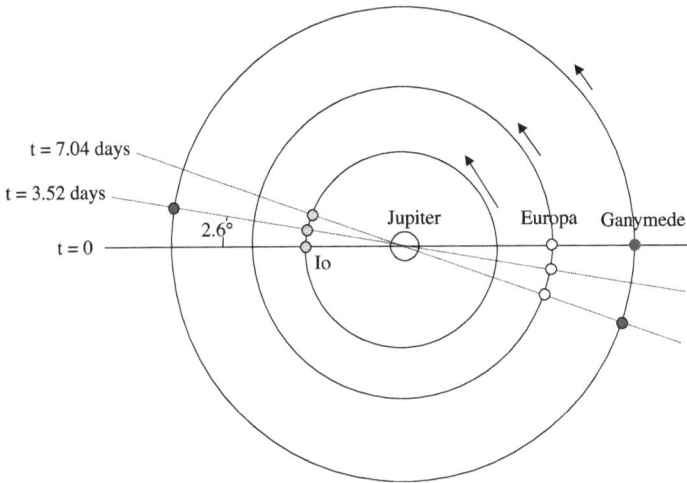

Figure 4.16 The possible alignments of the three inner Galilean satellites. The orbits and Jupiter are drawn to scale but, for clarity, the angles are exaggerated. The relative arrangement is repeated every 7.04 days.

Ganymede respectively. In addition the three satellites can never line up on the same side of Jupiter; the allowed conjunctions and oppositions are shown in Figure 4.16. There are two different relative positions when the satellites are in a line, with a repeat period of 7.04 days. One is where Io is on the opposite side of Saturn to Europa and Ganymede and the other where Europa is opposed to Io and Ganymede.

A general explanation for the approximate 1:2:4 resonance was suggested by C.F. Yoder in 1979. An important mechanism in his explanation was the same as that by which the tidal influence of the Moon on the Earth causes the Moon to gain energy and slowly recede from the Earth. Figure 4.17 shows high tides on the Earth in

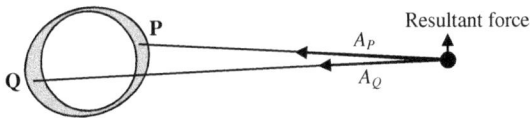

Figure 4.17 The resultant force on the Moon due to the displaced tidal bulges on Earth.

both the general direction of the Moon and on the opposite side of the Earth. Due to the Earth's spin these tides are dragged forward and appear at P and Q as shown in the figure. The accelerations produced on the Moon due to these displaced tidal bulges are shown as A_P and A_Q but, since P is closer to the Moon, A_P is larger and the general effect is to produce a force on the Moon in its direction of motion. This increases the energy of its orbital motion, so causing it to retreat from the Earth.

The effect of Jupiter on its satellites is similar to that of the Earth on the Moon. The effect is strongest at *perijove*, the closest point on the orbit to Jupiter. If the effect *only* occurred at perijove then the new orbit would have the same perijove distance but the semi-major axis would increase (i.e. the eccentricity, e, would increase, as does that of the Moon around the Earth). However, the greater the eccentricity is, the greater the dissipation of energy in Io's orbit due to tidal flexing. A stable equilibrium is established where the eccentricity induced by Jupiter and the other satellites gives a loss of energy that just balances the input of energy by Jupiter. If the period of Io, P_{Io}, were below $1/2\,P_{\text{Europa}}$, then Jupiter's tidal influence, acting most strongly on Io, would have pushed it outwards until its period was just below $1/2\,P_{\text{Europa}}$. The effect of Europa, as explained in Section 4.3, would then be to increase Io's eccentricity until the resultant loss of energy prevented any expansion of Io's orbit relative to Europa. At this stage the tidal effects on both Io and Europa would drive out the coupled satellites, Io and Europa, until P_{Europa} was just less than $1/2\,P_{\text{Ganymede}}$. In principle the three coupled satellites could be driven out to link up with Callisto but in practice the Solar System will not last long enough for this to happen.

4.4.3   *The commensurabilities of some of Saturn's satellites*

There are a number of commensurabilities involving Saturn's satellites, but not necessarily linking those that are adjacent in their

Table 4.3 Saturn's satellites involved in commensurability relationships.

Satellite	Diameter (km)	Period (days)
Mimas	397	0.9422
Enceladus	504	1.3702
Tethys	1006	1.8878
Telesto	28	1.8878
Calypso	31	1.8878
Dione	1123	2.7369
Helene	33	2.7369
Polydeuces	3	2.7369
Titan	5151	15.9452
Hyperion	292	21.2766

orbits. There are 61 satellites known for Saturn; in Table 4.3 there are listed the satellites involved in commensurabilities, their diameters, and the period of their orbits.

The commensurabilities shown by this table are:

$$\text{Tethys:Mimas} = 2.004{:}1 \approx 2{:}1$$

$$\text{Dione:Enceladus} = 1.997{:}1 \approx 2{:}1$$

$$\text{Hyperion:Titan} = 1.334{:}1 \approx 4{:}3$$

$$\text{Tethys:Telesto:Calypso} = 1{:}1{:}1$$

$$\text{Dione:Helene:Polydeuces} = 1{:}1{:}1.$$

The first three commensurabilities, involving pairs of satellites, can be explained by the mechanism that was suggested by Yoder for the Galilean satellites. However, the final pair of commensurabilities, connecting three satellites all with identical periods, involves a completely different mechanism that is fully explained in Section 4.4.4.

Exercise 4.5 The second largest satellite of Saturn is Rhea (diameter 1527 km). Its orbital period is 4.5182 days. Find possible commensurabilities with the satellites listed in Table 4.1.

4.4.4 *The Trojan asteroids*

There are two groups of asteroids, called the *Trojans*, which move in Jupiter's orbit, one group 60° behind Jupiter and the other group leading Jupiter by 60°. As a rule, an analytical solution is only possible for the motion of two bodies under mutual gravitational attraction. However, there are special three-body problems that can be solved where there are two bodies of finite mass plus one body of effectively zero mass. The problem of the Trojan asteroids is one such and we now show that the three bodies — the Sun, Jupiter and a Trojan asteroid — can be in equilibrium in circular orbits about the centre of mass of the system.

In Figure 4.18a, S and J represent the Sun and Jupiter, of masses $M_\odot$ and M_J respectively. They both orbit around the centre of mass, C, and for circular orbits the angular velocity is given by

$$\omega^2 = \frac{G(M_\odot + M_J)}{R^3}, \tag{4.13}$$

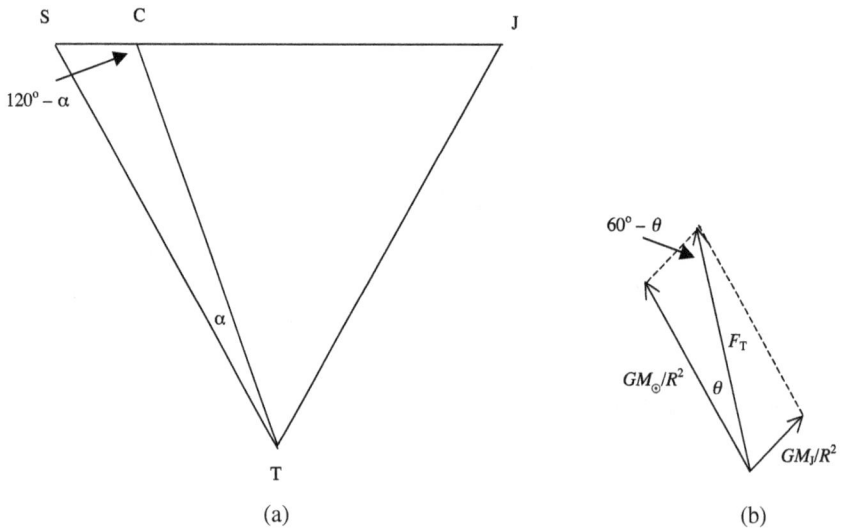

Figure 4.18 (a) The relative positions of the Sun (S), Jupiter (J), the Trojan asteroid (T) and the centre of mass (C). (b) The forces on the asteroid due to the Sun and Jupiter and the resultant F_T.

where R is the Sun–Jupiter distance. The Trojan asteroid, T, of negligible mass, is positioned so that SJT forms an equilateral triangle. For the asteroid to orbit in equilibrium, it too must move in a circular orbit about C with the angular velocity indicated by (4.13).

We now show that the combined gravitational fields of the Sun and Jupiter at T point along direction TC. In the triangle STC, $ST = R$ and $SC = M_J R/(M_J + M_\odot)$. Hence we can find a relationship involving the angle α (in degrees) as

$$\frac{SC}{\sin \alpha} = \frac{ST}{\sin(120 - \alpha)}$$

or

$$\frac{M_J}{\sin \alpha} = \frac{M_\odot + M_J}{\sin(120 - \alpha)}.$$

This reduces to

$$M_J \{\sin(120 - \alpha) - \sin \alpha\} = M_J \sin(60 - \alpha) = M_\odot \sin \alpha^1. \quad (4.14)$$

In Figure 4.18b the fields at T due to the Sun and Jupiter are shown; since the asteroid is equidistant from the other two bodies, these fields are proportional to $M_\odot$ and M_J respectively. The resultant field makes an angle θ with the direction TS given by

$$\frac{M_J}{\sin \theta} = \frac{M_\odot}{\sin(60 - \theta)}. \quad (4.15)$$

Comparing (4.14) and (4.15), it is clear that $\alpha = \theta$ so that the net field at T points towards the centre of mass of the system. The magnitude of the field, from Figure 4.18b, is

$$F_T = \frac{G}{R^2} \left(M_\odot^2 + M_J^2 + M_\odot M_J\right)^{1/2}. \quad (4.16)$$

---

[1] Here we have used the relationship $\sin A - \sin B = 2 \sin \left(\frac{A-B}{2}\right) \cos \left(\frac{A+B}{2}\right)$.

The distance TC $(= a)$ is found from the triangle SCT as

$$a^2 = R^2 + \frac{M_J^2}{(M_\odot + M_J)^2}R^2 - \frac{M_J}{M_\odot + M_J}R^2$$

$$= R^2 \frac{M_\odot^2 + M_J^2 + M_\odot M_J}{(M_\odot + M_J)^2}. \tag{4.17}$$

For an orbit of radius a with the angular velocity given by (4.13) the required field is

$$a\omega^2 = R\frac{\left(M_\odot^2 + M_J^2 + M_\odot M_J\right)^{1/2}}{M_\odot + M_J}\frac{G\left(M_\odot + M_J\right)}{R^3}$$

$$= \frac{G}{R^2}\left(M_\odot^2 + M_J^2 + M_\odot M_J\right)^{1/2}, \tag{4.18}$$

which is the field given in (4.16).

What has been shown is that the system of three bodies, rotating about the centre of mass all at the same angular velocity, is in a state of equilibrium — but not necessarily in a state of *stable* equilibrium. To show the stability of the three-body system involves a straightforward but lengthy analysis; however, we can show the stability numerically. Appendix III lists a FORTRAN program, TROJANS, which calculates the motion of an asteroid relative to the Sun and Jupiter. The calculation is relative to the Sun at the origin. By a suitable transformation, the position of the asteroid is found relative to the Sun–Jupiter line as the y-axis. As for the program given in Appendix II, a system of units is used where the unit of mass is the solar mass, the unit of distance is the astronomical unit (au) and the unit of time is one year. In such a system the gravitational constant is $4\pi^2$. Table 4.4 shows the input data that are requested by the program. The Sun, of unit mass, is stationary at the origin. Jupiter, with mass 0.001 units, is at distance 5.2 au on the y-axis moving in the negative x direction with a speed 2.75674 au yr^{-1}. The coordinates and velocity components for the trailing and leading asteroids can be derived from geometry. In fact, to illustrate the stability of the system the asteroid positions and velocities are not precisely those required for motion in a fixed position. The computed motions of both the leading

Table 4.4 The initial parameters for TROJAN.FOR. The position and velocity components of the asteroids are heavily rounded off to demonstrate the stability by oscillation around a mean position.

	Sun	Jupiter	Trojan 1	Trojan 2
Mass	1	0.001	0	0
x	0	0	−4.50	4.50
y	0	5.2	2.6	2.6
z	0	0	0	0
V_x	0	−2.75674	−1.38	−1.38
V_y	0	0	−2.39	2.39
V_z	0	0	0	0

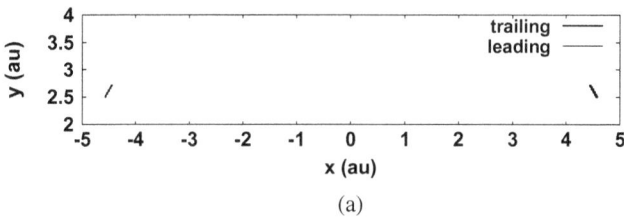

(a)

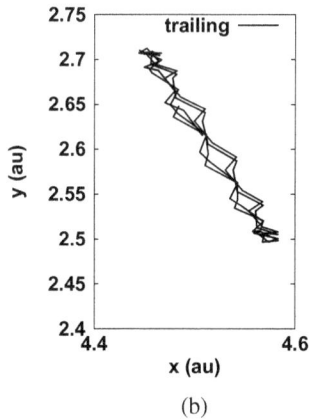

(b)

Figure 4.19 (a) Motions of the leading and trailing Trojan asteroid relative to the Sun and Jupiter. (b) Details of the motion of the trailing asteroid.

and trailing asteroids are shown in Figure 4.19 for a simulated period of 300 years, about 25 Jupiter orbits. The asteroids wander around the equilibrium position, showing clearly the stability of the system. The observed Trojan asteroids occupy a considerable region around the 60° leading and trailing positions, and their motions relative to Jupiter must resemble those shown in Figure 4.19.

The Saturn triplets of satellites Tethys–Telesto–Calypso and Dione–Helene–Polydeuces have the same Trojan-asteroid configuration. Saturn plays the role of the Sun as the central body, Tethys and Dione, comparatively massive satellites, respectively play the role of Jupiter for the two systems. Telesto is 60° ahead of Tethys in its orbit and Calypso 60° behind and, in the other system, Helene is 60° ahead of Dione and Polydeuces 60° behind.

It is not certain how these systems originated and some of the forces involved are very small but one has to remember that the Solar System is 4.5 billion years old so these systems have plenty of time within which to become established.

Problems 4

4.1 The Moon, with radius 1.738×10^6 km and in an orbit of mean radius 3.848×10^8 km and eccentricity 0.056, is made of similar material to Io. Calculate the rate at which energy is generated within it due to tidal flexing.

4.2 The mass of Saturn is 5.685×10^{26} kg. The Saturn satellite Tethys has mass 7.55×10^{20} kg and is in a circular orbit of radius 2.95×10^5 km. The tiny satellites Telesto and Calypso are in the same orbit, one approximately 60° ahead of Tethys and the other 60° behind.

 (i) What is the speed in au year^{-1} of Tethys in its orbit around Saturn?

 (ii) Placing Saturn at the origin and Tethys on the y axis then what are the (x, y) coordinates of Telesto and Calypso in au?

(iii) What are the velocity components in units of au yr^{-1} of Telesto and Calypso at these points?

Make a table, similar to Table 4.3 for data input to the program TROJANS. Those with the facilities to do so, run TROJANS. Recommended parameters are:

Initial timestep 0.01 (yr)

Total simulation time 1 000 (yr)

Tolerance 10^{-8} (au)

In view of the small mass of Tethys relative to Saturn, the oscillations of the small satellites are large and four-figure accuracy should be used in specifying their positions and velocities.

The output files for the leading and trailing satellites are in FAST.DAT and LAST.DAT respectively and these can be used to graphically display the output.

(The speed of a body of negligible mass in a circular orbit of radius r round a central body of mass M is $(GM/r)^{1/2}$.)

Chapter 5

Nuclear Magnetic Resonance

5.1 A Brief Review of the Structure of Atoms

There are two basic components of an atom, the nucleus and the accompanying electrons. The nucleus is very compact, with a diameter of the order of femtometres (fm, 10^{-15} m), and in general contains both protons and neutrons of approximately equal mass but with the proton having a positive charge while the neutron is electrically neutral — hence its name. The exception to a nucleus containing both protons and neutrons is the hydrogen atom, the nucleus of which contains a single proton only.

Electrons have a negative charge, equal in magnitude to that of the proton, and since there are as many of them in an atom as there are protons in the nucleus, the complete atom is electrically neutral. The electrons occupy a comparatively large region around the nucleus, giving the whole atom an effective diameter from a fraction of to a few times 10^{-10} m, depending on the number of electrons. Although in some circumstances electrons can have an identity as particles, with a mass that is $1/1836$ that of a proton, the state of their existence around the nucleus is properly described only by quantum mechanics. Each electron has an associated *state function*, which can be interpreted in different ways. Its square — or more strictly its modulus squared since it can be a complex quantity — can be interpreted either as a *continuous distribution of electron density* occupying all the space around the nucleus or as a *probability density* describing the probability per unit volume of finding the electron at any position. The state function defines all the properties of the electron; in particular, each electron has an associated orbital angular momentum which is of the form $J = l\hbar$, where l is a non-negative integer, $\hbar = h/2\pi$ and h is Planck's constant, 6.626×10^{-34} J s.

The chemical nature of an atom depends on the number of protons in the nucleus, or perhaps one should say the number of electrons because it is the interaction of electrons of different atoms that form molecules (Section 6.1.2). For all chemical elements there are *isotopes* having different numbers of neutrons, some of them stable and others radioactive. A carbon atom has six protons, and the most common isotope has six neutrons giving twelve units of mass in the nucleus. This can be represented as $^{12}_{6}$C, the C representing carbon, the 6 the number of protons, known as the *atomic number* and represented by the symbol Z (actually redundant since carbon *must* have six carbons) and the 12 the total number of nucleons (protons and neutrons) in the nucleus, known as the *mass number* and indicated as A. However, there is another stable isotope, $^{13}_{6}$C, with 7 neutrons and it accounts for about 11% of the carbon on Earth. There are many other isotopes of carbon that are radioactive, one of which, $^{14}_{6}$C, has a long half-life of 5,730 years and is of importance in the dating of ancient artefacts. For some atoms there is only a single stable isotope (e.g. sodium-23, $^{23}_{11}$Na, and aluminium-27, $^{27}_{13}$Al). As an extreme case of a chemical element with many stable isotopes there are ten for tin (chemical symbol Sn):

$$^{112}_{50}\text{Sn} \quad ^{114}_{50}\text{Sn} \quad ^{115}_{50}\text{Sn} \quad ^{116}_{50}\text{Sn} \quad ^{117}_{50}\text{Sn} \quad ^{118}_{50}\text{Sn} \quad ^{119}_{50}\text{Sn} \quad ^{120}_{50}\text{Sn} \quad ^{122}_{50}\text{Sn} \quad \text{and} \quad ^{124}_{50}\text{Sn}.$$

Exercise 5.1 Find the number of protons and neutrons in the following isotopes of uranium, lead and samarium: $^{235}_{92}$U, $^{238}_{92}$U, $^{204}_{82}$Pb, $^{154}_{62}$Sm.

5.2   Intrinsic Spins and Magnetic Moments

The three basic particles — proton, neutron and electron — have an intrinsic spin with a magnitude $1/2\,\hbar$; particles with half-integral intrinsic spin are collectively known as *fermions*. They are subject to *Fermi–Dirac statistics*, which governs the number density of particles in states with different energies. Associated with spin there is a magnetic moment, so these half-integral spin particles behave like tiny magnets, each with a magnetic moment that depends on the properties of the particles and so is different for electrons, neutrons and protons.

In a nucleus containing many nucleons the magnetic dipole axes of the proton and neutrons are collinear — some with north poles pointing in one direction and others in the opposite direction — and there is a tendency for them to form pairs with opposite polarities. The vast majority of nuclei with both an even number of protons and an even number of neutrons have zero net spin, suggesting that proton pairs with opposite spin and neutron pairs with opposite spin have formed to give a net zero spin. However, in general there will be a net spin for the whole nucleus. If the nucleus contains an even number of nucleons then there will be integral spin ($I\hbar$ with $I = 0, 1, 2$ etc.), while an odd number of nucleons will give half-integral spin ($I\hbar$ with $I = \frac{1}{2}, \frac{3}{2}, \frac{5}{2}$ etc.). Associated with this net spin will be a net magnetic moment. As an example of how this works, in Table 5.1 there are given the net nuclear spins, indicated by the symbol I, for various isotopes of iron. The value of Z, 26, shows that there are an even number of protons and where A is even there are also an even number of neutrons, giving the conditions for zero net spin.

For any particular particle, or system of particles, the *gyromagnetic ratio*, γ, is defined by

$$\gamma = \frac{\text{magnetic dipole moment}}{\text{angular momentum}}. \tag{5.1}$$

Table 5.1 The nuclear spin, I, for isotopes of iron.

Z	A	I
26	54	0
26	55	3/2
26	56	0
26	57	1/2
26	58	0
26	60	0

Thus for a nucleus with angular momentum $I\hbar$ the magnetic dipole moment is

$$\mu = \gamma I \hbar. \tag{5.2}$$

We now consider the dimensions of the quantities present in (5.1). Dipole moment is magnetic pole strength × distance and force is magnetic pole strength × magnetic field. The SI unit of magnetic field is the *tesla* (T), so putting these together gives the dimensions of dipole moment as kg m^2 s^{-2} T^{-1}. Angular momentum is mass × length^2 × angular velocity giving

$$\text{dimension}(\gamma) = \frac{\text{kg m}^2 \text{ s}^{-2} \text{ T}^{-1}}{\text{kg m}^2 \text{ s}^{-1}} = \text{s}^{-1} \text{ T}^{-1}.$$

Actually, the units in which γ is usually expressed include the dimensionless angular measure of the radian so that the dimension of γ is given as rad s^{-1} T^{-1}. The reason for this will become apparent in Section 5.3.1.

Exercise 5.2 Find the magnetic dipole moment of the nucleus of the isotope $^{19}_{9}$F, with spin $\frac{1}{2}\hbar$, given that the value of γ is 2.517×10^8 rad s^{-1} T^{-1}.

5.2.1 *Orientation of nuclei in a magnetic field*

With no external magnetic field present, the nuclear magnets orient themselves in random directions. A compass-needle magnet always aligns itself with the Earth's magnetic field and points towards the Earth's magnetic north pole but the nuclear magnets, when in a steady magnetic field, behave rather differently. They take up a finite number of configurations, with the components of the spin angular momentum in the direction of the field differing from each other by $\hbar$ and varying from $+I\hbar$ to $-I\hbar$. The six possible orientations for $I = 5/2$ are shown in Figure 5.1. For a general I, the number of different orientations is $2I + 1$.

For the example shown in Figure 5.1 the components of angular momentum are $m\hbar$ with $m = 5/2, 3/2, 1/2, -1/2, -3/2$ and $-5/2$.

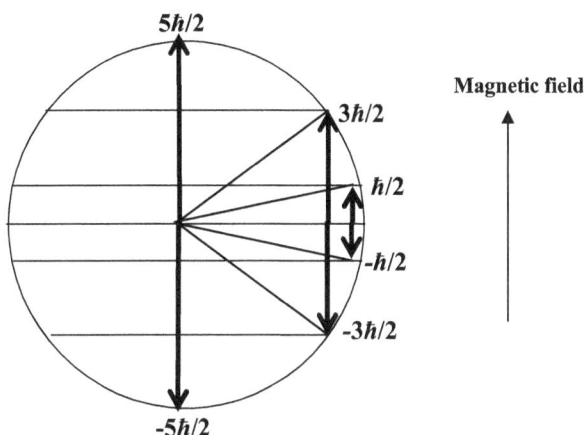

Figure 5.1 The six components of spin angular momentum in the direction of a magnetic field for I = 5/2.

The value of m, the *nuclear magnetic spin quantum number,* gives the component of magnetic moment in the direction of the field.

It can be shown that for any rotating charged body in which the local charge density is proportional to the local mass density then, from a classical point of view, the gyromagnetic ratio is given by

$$\gamma = \frac{q}{2m} \times \text{angular momentum}, \tag{5.3}$$

in which q is the total charge on the body and m its mass. Quantum mechanics, which is appropriate for atomic and sub-atomic scale systems, gives a different result but with the same dependence on the properties of the system. The magnetic moments of nucleons or nuclei are given in units of the *nuclear magneton* defined by

$$\mu_N = \frac{e\hbar}{2m_p} = 5.0508 \times 10^{-27} \text{J T}^{-1}, \tag{5.4}$$

in which e is the proton charge and m_p is the mass of the proton. The dipole magnetic moment of any nucleon or nucleus is proportional to I and is given by

$$\mu = g\mathrm{I}\mu_N \tag{5.5}$$

and its component of magnetic moment in the direction of the field is

$$\mu_m = gm\mu_N = \gamma m\hbar. \tag{5.6}$$

The quantity g is the Landé g-factor, first introduced by the German, later American, physicist Alfred Landé (1888–1976). For the proton, g is 5.5857 and for the neutron it is -3.8261. Since we have related the magnetic moment of a particle to its charge it might seem rather strange that a neutron should have a magnetic moment. This apparent anomaly is bound up with the sub-structure of the neutron in terms of *quarks*, which are also fermions with charges and magnetic moments. A neutron is made up of two similar *down quarks*, each with a charge $-\frac{1}{3}e$, and an *up quark*, with a charge $+\frac{2}{3}e$. Although the charges balance to give a neutral particle the dipole magnetic moments do not (they have different g-factors), thus giving the neutron a net magnetic moment.

Every orientation of a magnetic dipole in a magnetic field has associated with it an energy that depends on m and the strength of the magnetic field, B. This energy is given by

$$E_m = -\mu_m B, \tag{5.7}$$

from which it is clear that for positive m, corresponding to a positive component of magnetic moment in the direction of the field, the energies are negative.

At this stage we call on a result from statistical mechanics, the *Boltzmann distribution*, which states that for a system of particles at absolute temperature T in equilibrium, with a number of possible states, the nth one of which has energy E_n, the number with that energy is given by

$$N_n = Cg_n \exp\left(-\frac{E_n}{kT}\right), \tag{5.8}$$

where C is a constant, k is Boltzmann's constant, $1.381 \times 10^{-23}\ \mathrm{J\,K^{-1}}$, and g_n is the *degeneracy* of the energy level (i.e. the number of different states with that energy). Using this result for an ensemble of nuclei, the relative numbers with particular values of m will be given by the Boltzmann distribution so the actual proportion will be

$$P(m) = \frac{\exp\left\{-E_m/(kT)\right\}}{\sum\limits_{j=-I}^{I} \exp\left\{-E_j/(kT)\right\}}, \tag{5.9}$$

giving $\sum\limits_{m=-I}^{I} P(m) = 1.$

Exercise 5.3 What are the proportions of spin-up and spin-down states of $^{19}_{9}\text{F}$ in a magnetic field of intensity $3\,\text{T}$ at a temperature of $100\,\text{K}$? (See Exercise 5.2 for information about $^{19}_{9}\text{F}$.)

5.3 Magnetic Resonance Imaging (MRI)

We are now going to describe a resonance phenomenon associated with the magnetism of nuclei: *nuclear magnetic resonance* (NMR). The best-known application of NMR is in the field of medicine where it is known as *magnetic resonance imaging* (MRI), a means of imaging the internal structure of the body that does not involve exposure to potentially harmful X-radiation. It depends on the fact that hydrogen, a major component of the body, mainly as a constituent of water, appears in varying amounts in different body materials; by locating the positions of the protons — the nuclei of hydrogen — an image is obtained giving contrast between different kinds of body tissue.

The basic mechanism of MRI depends on the behaviour of the proton in a magnetic field. For the proton $I = \frac{1}{2}$ so the proton can take up one of two configurations with the magnetic dipole either parallel or anti-parallel to the field, corresponding to $m = \frac{1}{2}$ and $m = -\frac{1}{2}$. The ratio of the probabilities of the two orientations, from Equations (5.6), (5.7) and (5.8), is given by

$$\frac{P_{1/2}}{P_{-1/2}} = \frac{\exp\left(\frac{\frac{1}{2}g\mu_N B}{kT}\right)}{\exp\left(-\frac{\frac{1}{2}g\mu_N B}{kT}\right)}. \tag{5.10}$$

For a field of $1\,\mathrm{T}$ and a temperature of $300\,\mathrm{K}$

$$\frac{\frac{1}{2}g\mu_N B}{kT} = \frac{0.5 \times 5.5857 \times 5.0508 \times 10^{-27} \times 1}{1.381 \times 10^{-23} \times 300} = 3.405 \times 10^{-6}.$$

This value is much less than unity so we can approximate (5.10) to

$$\frac{P_{1/2}}{P_{-1/2}} = \frac{1 + \frac{\frac{1}{2}g\mu_N B}{kT}}{1 - \frac{\frac{1}{2}g\mu_N B}{kT}}.$$

Dividing top and bottom by 2 gives

$$\frac{P_{1/2}}{P_{-1/2}} = \frac{0.5 + \beta}{0.5 - \beta} \tag{5.11}$$

where

$$\beta = \frac{g\mu_N B}{4kT}.$$

Since the sum of the numerator and divisor in (5.11) is 1, the numerator and divisor are $P_{1/2}$ and $P_{-1/2}$, respectively. This gives the proportion of parallel dipoles as 0.500001702, a very slight preponderance over the number in the anti-parallel configuration, but it is sufficient to enable the MRI system to produce images.

5.3.1 *Larmor precession*

When a magnetic dipole is in a magnetic field it experiences a torque, as illustrated in Figure 5.2a, which shows the dipole as a positive magnetic pole of strength p and a negative magnetic pole of equal magnitude, separated by a distance d, giving a dipole moment

$$\mu = pd. \tag{5.12}$$

The torque exerted on the dipole by the field is

$$\tau = pdB\sin\theta = \mu B\sin\theta. \tag{5.13}$$

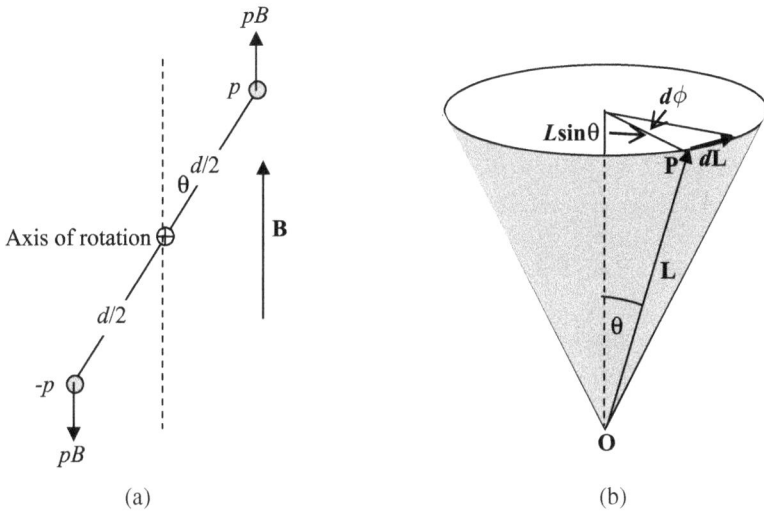

Figure 5.2 (a) Torque and rotation axis due to magnetic field and dipole. (b) Precession of the magnetic dipole due to the field.

The dipole and the field are in the plane of the figure and the angular motion induced by the torque is about an axis perpendicular to the plane of the figure; the vectors representing both the torque and the angular momentum it induces will be along that axis, perpendicular to the plane defined by the dipole and the field.

In Figure 5.2b the angular momentum associated with the dipole is represented by the vector $\mathbf{L}$ (magnitude L), which must be of the form $I\hbar$, and the angular momentum added by the torque exerted on the dipole in a time dt is indicated as $d\mathbf{L}$. This is perpendicular to both $\mathbf{B}$ and $\mathbf{L}$ and is tangential to a cone with axis parallel to $\mathbf{B}$ that has $\mathbf{L}$ on its surface. The effect of this is to move the vector representing the dipole so that OP traces out the surface of the shaded cone. The angular momentum vector, and the associated parallel vector representing the dipole moment, undergoes *Larmor precession*, named after the Irish physicist, Joseph Larmor (1857–1942), who first described the effect.

From (5.5) the dipole magnetic moment is

$$\mu = gI\mu_N = gL\mu_N/\hbar. \tag{5.14}$$

The torque gives the rate of change of angular momentum so that, with $d\phi$ as shown in Figure 5.2b and from Equations (5.13) and (5.14),

$$\tau = \mu B \sin\theta = \left|\frac{d\mathbf{L}}{dt}\right| = \frac{L\sin\theta d\phi}{dt} = gL\mu_N \mathbf{B}\sin\theta/\hbar$$

and that gives the Larmor precession rate

$$\omega_L = \frac{d\phi}{dt} = \frac{g\mu_N B}{\hbar} = g\frac{eB}{2m_p}. \tag{5.15}$$

For a proton in a field of 1 tesla,

$$\omega_p = 5.5857 \times \frac{1.602 \times 10^{-19} \times 1}{2 \times 1.673 \times 10^{-27}} = 2.674 \times 10^8 \, \text{s}^{-1}$$

and the *Larmor frequency* is

$$\nu_p = \frac{\omega_p}{2\pi} = 42.56 \, \text{MHz},$$

which is in the radiofrequency region of the electromagnetic spectrum.

From (5.2) and (5.5) we have

$$\mu = \gamma I\hbar = g\mu_N I$$

giving

$$\gamma = \frac{g\mu_N}{\hbar}$$

so that, from (5.15)

$$\omega_p = \gamma B \quad \text{and} \quad \nu_p = \frac{\gamma B}{2\pi}, \tag{5.16}$$

an alternative expression for the Larmor frequency in terms of the gyromagnetic ratio.

Exercise 5.4 What is the Larmor frequency for a $^{19}_{9}$F nucleus in a magnetic field of intensity 2 T? (See Exercise 5.2.)

5.3.2 *The basic physics of the MRI process*

As previously mentioned, the working of MRI depends on the presence of large amounts of hydrogen (and hence protons) in water, a major component of the tissues of the body. The two orientations of the dipole moment of a proton with a magnetic field have different energies, the energy of the anti-parallel configuration, μB, being higher than that of the parallel configuration, $-\mu B$.

We have implied that there are only two possible orientations of the proton dipole with a magnetic field; that means that if we were to carry out an experiment to *observe* the orientation of a proton then it would only be found in one or other of these orientations. In the language of quantum mechanics these are *eigenstates* of the entity — the only ones that can be observed. However, quantum mechanics does allow an *unobserved* entity, for example a proton in a magnetic field, to be in a *mixed state*, a mixture of eigenstates. So, for example, if there were two possible eigenstates, represented by u_1 and u_2, a mixed state would be expressed as

$$\psi = c_1 u_1 + c_2 u_2, \tag{5.17}$$

under the condition that $c_1^2 + c_2^2 = 1$. If an experiment were then made on the entity to determine its state, the probability that it would be found in states represented by u_1 and u_2 would be c_1^2 and c_2^2, respectively. If there is a large ensemble of entities (e.g. many protons), then an alternative description of the ensemble is that a proportion c_1^2 was in state u_1 and c_2^2 in state u_2; the experiment that measured the states of many component entities of the ensemble could not distinguish the two cases.

If a radio-frequency electromagnetic field with *precisely* the Larmor frequency is applied to the protons then some parallel dipoles absorb energy. Individual dipoles tilt away from the parallel configuration and begin to move over towards the higher-energy anti-parallel direction; they are in a mixed state. Since the Larmor frequency is independent of the tilt, the resonance persists as the dipoles tilt and they can continue to absorb energy. The final extent of the tilting would depend on both the intensity and duration of the radio wave.

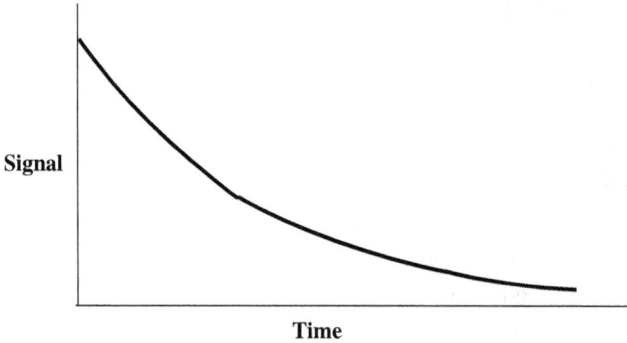

Figure 5.3 A declining signal with time after the radio-frequency excitation is removed.

Now, if the radio-frequency field is switched off then the input of energy ceases and the ensemble of dipoles is no longer in an equilibrium configuration with the proportions of the two states as given by (5.10). The tilted dipoles begin to return to the parallel configuration to give the original equilibrium proportion of parallel and anti-parallel dipoles. As they do so, electromagnetic radiation at the Larmor frequency is released. Some of it is absorbed by the surrounding tissue but most emerges from the body and provides the signal for the MRI image. The output signal declines with time in an exponential form, as shown in Figure 5.3. This decay process has a characteristic timescale, $T1$, which is the time for the signal to fall to $1/e$ of its original value. The signal at time t, $S(t)$, is related to that at time zero, $S(0)$, by

$$S(t) = s(0) \exp(-t/T1). \qquad (5.18)$$

MRI can be carried out in many ways — to optimize particular features that are to be imaged, some procedures involve the use of multiple pulses of electromagnetic radiation — but here we will just deal with the most basic imaging technique, which, if not the quickest and most effective, is the easiest to understand and adequately describes the principles involved.

It is common to use what is known as a *90° pulse* of electromagnetic radiation, which energizes the spin-up protons to the extent that the longitudinal magnetization due to the proton dipoles (that

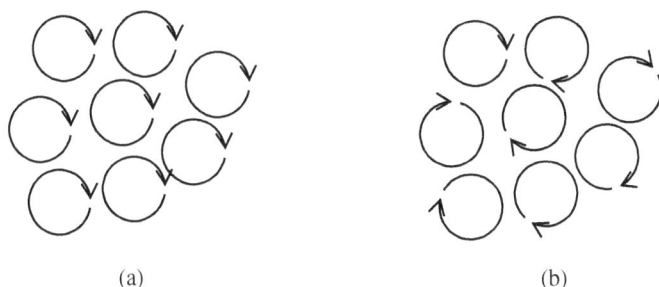

(a)   (b)

Figure 5.4 (a) Magnetic dipoles with precessions in phase. (b) Magnetic dipoles with precessions in random relative phases.

in the direction of the field) is zero, but still gives transverse magnetization — which is to say that the collection of dipoles gives a field perpendicular to the imposed magnetic field. When the radio-frequency pulse of energy is applied to the specimen all the dipoles undergo precession together. This state of affairs is illustrated in Figure 5.4a; the precessions are *in phase*. When the pulse is switched off the radio-frequency emission from each dipole is also in phase — that is to say the peaks and troughs of the radiation from each photon occur together and reinforce each other. The other feature of this system is that, in any direction perpendicular to the main field, the transverse magnetization will vary in a sinusoidal fashion and if a coil of wire is placed with its plane perpendicular to the transverse field a current will be induced in it that has the Larmor frequency.

In a collection of protons with differing spin orientations there is a process of *spin–spin interactions* taking place, where neighbouring protons exchange energy so that one goes from parallel to anti-parallel while the other does the reverse. Now, the magnetic dipoles not only exist in the externally applied magnetic field but also create a local magnetic field which can give an imbalance of spin-up and spin-down states. The random exchanges of energy between neighbouring protons give random fluctuations in local magnetic fields that, in their turn, change the local Larmor precession frequency. With some protons having slightly faster precessions than the average and some slightly slower, the rotations of different protons gradually become out of phase so that peaks and troughs of output from

Table 5.2 T1 and T2 values for some body tissues. (CSF is cerebrospinal fluid.)

Material	T1 (ms)	T2 (ms)
Muscle	870	47
Kidney	650	58
Grey matter	920	100
Lung	830	80
Liver	490	43
CSF	>3,000	~2,000

different dipoles are increasingly cancelling each other (Figure 5.4b), which will reduce the current through the coil. The characteristic timescale for this decay effect is T2, which also shows an exponential decline. The timescales for the T1 and T2 processes depend on the type of material involved and Table 5.2 gives some typical values.

Because there is variation of decay times (normally called *relaxation times*), if signals are compared from two regions containing different kinds of tissue then the relative strengths of the signals from the regions will vary with time. The measurement is of the cumulative signal from the beginning of the emission stage and the relative values from two regions will depend both on the density of hydrogen in the material and the time period for which the signal is collected. Typically, for good tissue discrimination a time of about 0.4 s is used. These accumulated signals, displayed on a CRT with intensity proportional to the total signal, show the difference in the two regions.

Now we need to describe the procedure that gives images from MRI. However, before we do that we need to understand the important mathematical tool by which the MRI signals are turned into an image: *Fourier transforms*.

Exercise 5.5 Tissue A, with $T1 = 500$ ms, has a hydrogen concentration 20% greater than tissue B, with $T1 = 900$ ms. What is the relative MRI signal after (i) 100 ms and (ii) 500 ms?

5.3.3 *Fourier transforms*

A comprehensive treatment of the Fourier transform (FT) would not be appropriate here and can be found in various mathematical text-books.[1] We will just concentrate on the main results that are relevant to the interpretation of MRI signals and express various equations in a form that can be immediately translated into MRI application. We take a distribution of density of a physical quantity represented by $\rho(x, y, z)$. The density can be of various kinds (e.g. the density of electrons in a crystal that is scattering X-rays or the strength of the MRI signal per unit volume coming from a point within a material). The FT of this distribution is given by

$$F(u, v, w) = \int\limits_{-\infty}^{\infty} \int\limits_{-\infty}^{\infty} \int\limits_{-\infty}^{\infty} \rho(x, y, z) \exp\left\{-2\pi i \left(xu + yv + zw\right)\right\} dxdydz$$

(5.19a)

and the *inverse Fourier transform* is

$$\rho(x, y, z) = \int\limits_{-\infty}^{\infty} \int\limits_{-\infty}^{\infty} \int\limits_{-\infty}^{\infty} F(u, v, w) \exp\left\{2\pi i \left(xu + yv + zw\right)\right\} dudvdw.$$

(5.19b)

The coordinates (x, y, z) are *real space coordinates* appropriate to the real physical system under investigation and (u, v, w) are *reciprocal space coordinates* corresponding to a hypothetical space that can be related geometrically to the real space.

We shall be concerned with the application of the FT in two dimensions for which the equations are

$$F(u, v) = \int\limits_{-\infty}^{\infty} \int\limits_{-\infty}^{\infty} \rho(x, y) \exp\left\{-2\pi i \left(xu + yv\right) dxdy\right\}$$

(5.20a)

---

[1] For example, M.M. Woolfson & M.S. Woolfson, Mathematics for Physics, Oxford University Press, 2007.

and

$$\rho(x, y) = \int\limits_{-\infty}^{\infty} \int\limits_{-\infty}^{\infty} F(u, v) \exp\left\{2\pi i \left(xu + yv\right) dudv\right\}. \qquad (5.20b)$$

What is clear from these equations is that if we could determine $F(u, v)$ for all u and v then we could determine the two-dimensional density $\rho(x, y)$. By writing

$$F(u, v) = \int\limits_{-\infty}^{\infty} \int\limits_{-\infty}^{\infty} \rho(x, y) \cos\left\{2\pi \left(xu + yv\right)\right\} dxdy$$

$$+ i \int\limits_{-\infty}^{\infty} \int\limits_{-\infty}^{\infty} \rho(x, y) \sin\left\{2\pi \left(xu + yv\right)\right\} dxdy = A + iB,$$

it is clear that $F(u, v)$ is a complex quantity with magnitude

$$|F(u, v)| = \sqrt{A^2 + B^2} \text{ and phase } \phi = \tan^{-1}(B/A). \qquad (5.21)$$

In Figure 5.5 we show a hypothetical distribution of density and its Fourier transform. Each point of the FT has a magnitude (illustrated by variable shades of grey) and a phase that can be anywhere

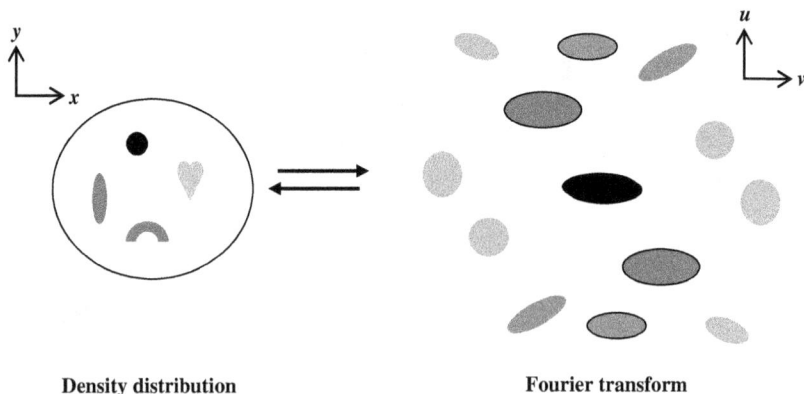

Density distribution   Fourier transform

Figure 5.5 A representation of a two-dimensional density distribution and its Fourier transform.

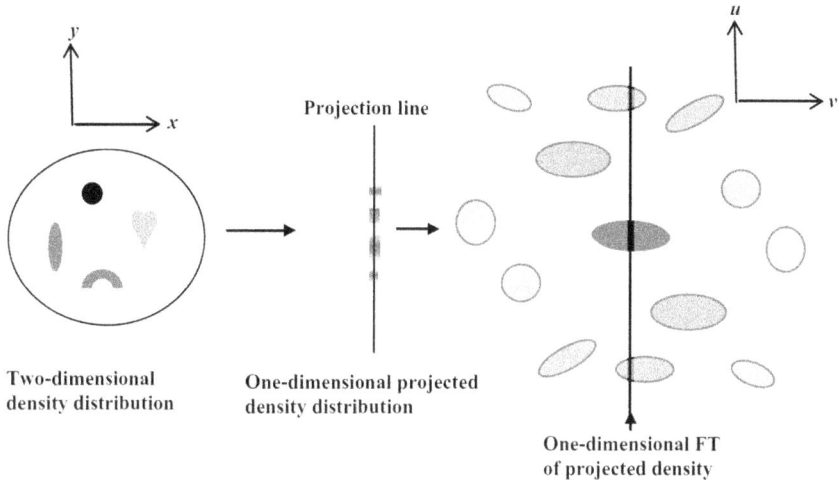

Figure 5.6 The one-dimensional FT of density projected onto a line.

in the range 0 to 2π. The arrows show that, by mathematical means, each can be derived from the other.

In Figure 5.6 we now show all the density projected onto a line. Since the projected density is one-dimensional, so is its FT. The essential basis of MRI is that its FT is just a *line of the two-dimensional FT*: that through its centre and parallel to the line on which the density is projected.

Exercise 5.6 For a one-dimensional Fourier transform $F(u) = \int_{-\infty}^{\infty} f(x) \exp(-2\pi i x u)dx$, find $F(0.1)$ and $F(-0.1)$ for $f(x) = 1$ for $-1 \le x \le 1$, and for $f(x) = 0$ otherwise.

5.3.4 *The MRI procedure*

We must now consider how spatial discrimination occurs in a MRI scan, which requires a description of the construction and operation of an MRI machine.

Figure 5.7 shows an MRI scanner, with a table on which the patient lies that can be inserted into a large tubular structure.

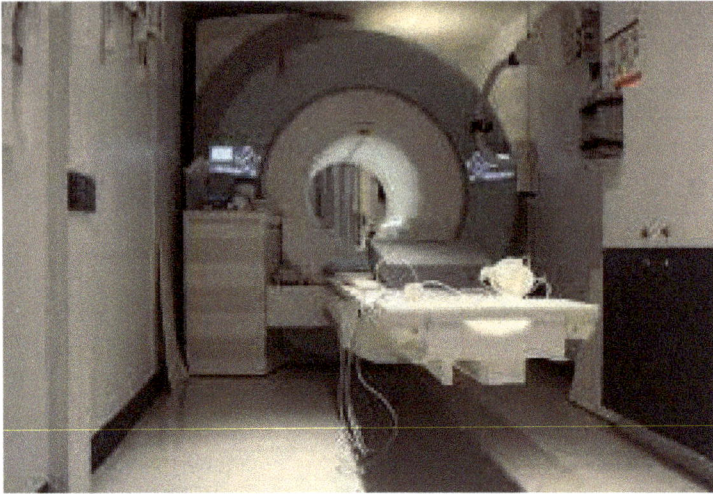

Figure 5.7 An MRI scanner.

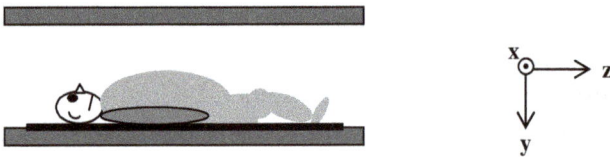

Figure 5.8 A schematic MRI scanner.

A schematic representation of the patient within the tube is shown in Figure 5.8.

Coils of wire are situated within the walls of the cylinder, and electric currents through these coils set up magnetic fields of various kinds within the cylinder. Firstly there are the *main coils* that give a large (0.2–3 T), almost uniform field throughout the length of the tube. For very high fields superconductor magnets are used, where the coils are cooled to liquid helium temperature so that their electrical resistance falls to zero and large currents can then be passed through them without excessive heating. Secondly, since it is not possible to produce a precisely uniform field with a single coil, there are also subsidiary *shim coils* that are used to modify the main field to make it as uniform as possible. Thirdly, there are three sets of *gradient coils* giving much smaller fields with a linear variation in the x-, y- and z-directions. The linear gradients are usually of the order $0.2 \, \text{mT m}^{-1}$.

We now describe how an xy scan would be carried out at a particular cross-section of the patient; this would be a section at constant z (e.g. a scan across the midriff). The stages in the process are:

(i) The main coils are switched on as well as the z-direction gradient coils, which gives a high but linearly varying field in the z-direction. Since B varies linearly along z then, from (5.13), so does the Larmor frequency. The patient is then exposed to a brief pulse of electromagnetic radiation at the Larmor frequency corresponding to the field at the z cross-section of interest; the proton dipoles at or very close to this cross-section will be the only ones affected. The range of z affected can be increased or decreased by reducing or increasing the field gradient along z.

(ii) At the conclusion of the exciting pulse the z-gradient is removed and simultaneously the x-gradient coils are switched on. This makes the Larmor frequency variable in the x-direction so that the frequency of the *emitted radiation* by the proton dipoles as they return towards equilibrium is x-dependent, but not y-dependent. During the period of about 0.4 s that the radio-frequency detector picks up the emitted radiation or measures the transverse field, it detects a range of frequencies, the intensity at each frequency being the combined contribution of the signal from the hydrogen content of the complete range of values of y at a particular value of x. This is equivalent to a projection of the image density, similar to what is shown in Figure 5.6.

(iii) A one-dimensional version of the FT (Exercise 5.6) is used to calculate a line in the two-dimensional FT of the complete two-dimensional image, both in magnitude and phase.

(iv) The whole process is now repeated with the modification that the x-gradient field is slightly reduced and a small y-gradient field is added so that the net field gradient in the xy plane is inclined at $1°$ to the x-axis. The FT of this intensity distribution

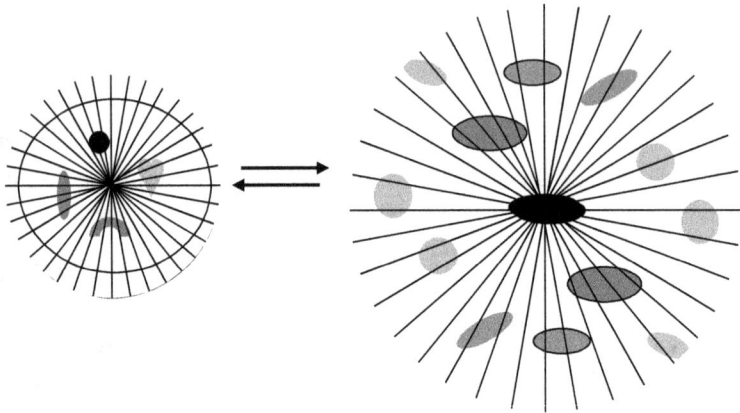

Figure 5.9 Projection lines in a two-dimensional density distribution and corresponding sampling lines of the two-dimensional transform.

is now a sample of the complete two-dimensional FT on a line making an angle of 1° with the original one.

(v) This process is repeated, each time changing both the x-gradient field and the y-gradient field to rotate the field gradient by 1° steps until the x-gradient field is the reverse of its initial value. In this way a sample of the complete two-dimensional FT is found.

(vi) The process is illustrated in Figure 5.9 (for clarity the projections are shown at 10° intervals). While there will be tiny gaps in a two-dimensional FT so found, the image obtained from the inverse Fourier transform gives a high-resolution image of the two-dimensional section of the patient; this is the MRI image.

The availability of producing field gradients in three dimensions gives great flexibility to the method so that one can, for example, obtain scans in cross-sections at any angle to the axes.

Variation of the collecting time of the signal makes it possible to emphasize either the T1 or T2 components from a particular type of tissue. Since the T1 and T2 data are stored in digital form

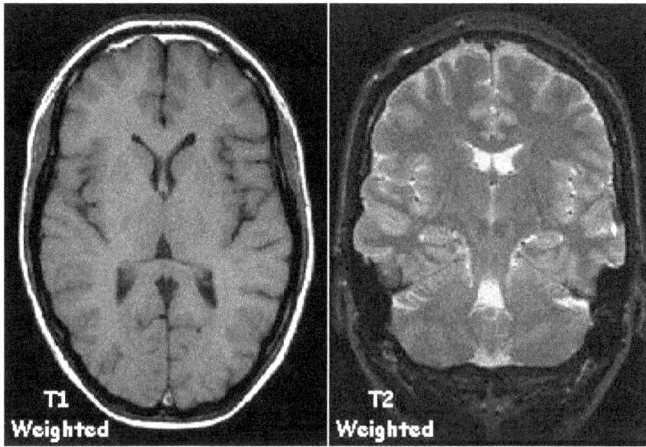

Figure 5.10 MRI brain scans taken with T1 and T2 signals.

they can be combined and manipulated in various ways to enhance particular features of interest. For example, one can produce an image for $T1 - nT2$ for any n, or $T2 - C$ where C is a constant. Figure 5.10 gives sample T1 and T2 brain scans that show the differences of contrast that can be obtained.

MRI has some disadvantages. The tube is narrow, which means that patients suffering from claustrophobia find it difficult to remain inside the machine for the 15 minutes required for a complete scan. It is also very noisy within the tube, which is also troublesome to some patients, although the provision of earphones and a choice of music can overcome that problem. Another problem is that one must ensure that no metal objects are present within the machine while it is running since they can be projected at high velocity when the large magnetic field is switched on and cause serious injury or even be fatal. Finally, it is unsuitable for use on any patient with a magnetic-metal implant or with a pacemaker, the working of which would be affected by the magnetic field. For such patients a CT (computed tomography) scan is used instead, which gives images of similar quality to those from MRI but with the disadvantage of exposure to (possibly) harmful X-radiation.

Exercise 5.7   The working region of an MRI scanner has a length of 2 m. If the main field is 2 T and a z-direction gradient of $2\,\text{mT m}^{-1}$ is superimposed, the additional z field being zero at one end, then what is the range of Larmor frequencies over the working region? (For a proton $\gamma = 2.6751 \times 10^8\,\text{rad s}^{-1}\,\text{T}^{-1}$.)

5.4   Other Applications of NMR

Although MRI is the aspect of NMR that most directly imacts everyday life, NMR also has important applications in many other areas of science, particularly in chemistry, biochemistry and materials science. All nuclei with dipole moments will have a characteristic Larmor frequency so that if a sample of material is excited by a wide range of radio-frequency radiation then the emitted NMR spectra give information about the types of atom present and their abundance. However, it can go further than that: the local magnetic field for a particular atom or molecule is affected by the arrangement of atoms in the immediate vicinity so that by careful analysis of the NMR spectrum structural information can be obtained. Although X-ray crystallography has played a leading role in the determination of the structures of biological macromolecules, NMR is now making an important contribution in the elucidation of such structures.

Problems 5

5.1 A nucleus with $A = 35$ and $I = 3/2$ has a g-factor of 5.722. If it is placed in a magnetic field of intensity 10 T then determine the energies of the possible orientations of the nuclear dipole and the probabilities of their occurrence, given a temperature 10 K.

5.2 Two neighbouring regions of a patient's anatomy, A and B, have the following characteristics:

Tissue	Relative proton density, ρ	T1 (ms)	T2 (ms)
A	1.00	500	80
B	0.95	900	60

What are the relative image intensities of the A and B regions for T1 images and T2 images with a total signal collecting time of (i) 0.4 s and (ii) 0.1 s?

Find the relative image intensities for $T1 - 3T2$ with collecting time 400 ms and for $T2 - 0.95 \times T2(B)$ with collecting time 100 ms.

Comment on your results.

Chapter 6

Electron Spin Resonance

6.1 The Electronic Structure of Molecules and Free Radicals

Since electrons are fermions they have half-integral spin with angular momentum magnitude $\frac{1}{2}\hbar$. For nuclei consisting of protons and neutrons the natural unit of magnetic moment is the nuclear magneton, given by Equation (5.4). Similarly, for the electron the unit of magnetic moment is the *Bohr magneton* given by

$$\mu_B = \frac{e\hbar}{2m_e},\tag{6.1}$$

where m_e is the electron mass. Since the electron is much less massive than the proton the Bohr magneton, with value $9.274 \times 10^{-24}\,\text{J·T}^{-1}$, is a much larger unit than the nuclear magneton.

6.1.1 *The electronic structure of atoms*

The electronic structure of an individual atom is governed by the *Pauli Exclusion Principle*, which states that no two of its electrons can have the same set of quantum numbers. The state function of an electron (Section 5.1) is generally described as its *orbital*, which is defined in terms of three quantum numbers, each associated with a different aspect of the orbital.

These are:

The Principal Quantum Number, n

This is a non-zero positive integer that describes the energy of the orbital. The closer an electron's negative charge is to the positively-charged nucleus the less is the energy associated with it. Higher principal quantum numbers correspond to orbitals that have electron density, on average, farther from the nucleus. However, all energies of bound electrons must be negative since a positive energy would correspond to an unbound electron.

The Azimuthal Quantum Number, l

This describes the angular momentum of the orbital, which is $l\hbar$ with the condition that $0 \leq l \leq n - 1$. For $l = 0$ the orbital is spherically symmetric. For higher values of l the orbitals have more complex symmetries and become more extended.

The Magnetic Quantum Number, m

This quantum number is best understood in terms of an applied magnetic field and describes the orientation of the orbital in the field. It must satisfy the condition $-l \leq m \leq l$. It is basically similar to what was described for integer nuclear spin in Section 5.2.1.

There is a fourth quantum number that is not involved in defining the orbital but which must be taken into account with respect to the Pauli Exclusion Principle.

The Electron-Spin Quantum Number, s

This is related to the intrinsic spin, and hence magnetic dipole moment, of the electron and again can best be understood in terms of reactions to a magnetic field. The possible values of the electron-spin quantum number are $+\frac{1}{2}$ and $-\frac{1}{2}$.

We can consider the electronic structure in terms of shells, corresponding to the value of n. Thus for $n = 1$ and $n = 2$ the possible sets of quantum numbers are given in Table 6.1. There are two electrons for $n = 1$ and 8 for $n = 2$ and in general the number is $2n^2$ (e.g. 18 for $n = 3$).

Table 6.1 The quantum numbers for $n = 1$ and $n = 2$.

n	l	m	s
1	0	0	$+\frac{1}{2}$
1	0	0	$-\frac{1}{2}$
2	0	0	$+\frac{1}{2}$
2	0	0	$-\frac{1}{2}$
2	1	1	$+\frac{1}{2}$
2	1	1	$-\frac{1}{2}$
2	1	0	$+\frac{1}{2}$
2	1	0	$-\frac{1}{2}$
2	1	-1	$+\frac{1}{2}$
2	1	-1	$-\frac{1}{2}$

Based on a now-defunct way of categorizing spectral lines, the lower-case letters s (sharp), p (principal), d (diffuse) and f (fundamental) are used to describe *subshells* for each value of n (i.e. s for $l = 0$, p for $l = 1$, d for $l = 2$ and f for $l = 3$). Subsequent subshells follow alphabetically (e.g. g for $l = 4$). Thus the electronic structure of carbon, with 6 electrons, can be described as $1s^2 2s^2 2p^2$ meaning that for $n = 1$ the s subshell contains 2 electrons (with opposite spins), for $n = 2$ the s subshell has 2 electrons and the p subshell has 2 electrons. For sodium, with eleven electrons, the structure is $1s^2 2s^2 2p^6 3s^1$. For normal stable atoms the shells and subshells fill up in order of energy with the lowest energy first. Although all electrons with the same value of n have *approximately* the same energy, there are differences (usually, but not always, small in energy) between subshells in the order $E_s < E_p < E_d < E_f$. Because of these differences the shells do not always fill up in the way that might seem logical. For example, for argon, with eighteen electrons, the structure is $1s^2 2s^2 2p^6 3s^2 3p^6$, which follows a logical progression. However, for the atom with the next higher atomic number, potassium with nineteen electrons, because of the extreme shape of their orbitals, the $3d$ electrons have higher energy than the $4s$ electrons so its electronic structure is $1s^2 2s^2 2p^6 3s^2 3p^6 4s^1$ and it is followed by calcium, with twenty electrons, with structure $1s^2 2s^2 2p^6 3s^2 3p^6 4s^2$. For the next element, scandium with twenty-one electrons, the lowest energy for the next electron is in subshell $3d$ so its electronic structure is $1s^2 2s^2 2p^6 3s^2 3p^6 3d^1 4s^2$, giving a lone electron in the 3d subshell.

Exercise 6.1   What is the electronic structure of (a) boron, $^{11}_{5}$B and (b) fluorine, $^{19}_{9}$F?

6.1.2   *The electronic structure of molecules*

For individual atoms with an even number of electrons (e.g. carbon), the spins occur in pairs with one spin-up and the other spin-down so that there is no net electron spin. However, for sodium with its lone $3s$ electron, there is a net spin and hence a net magnetic moment. The same is true for chlorine with seventeen electrons and an electronic configuration $1s^2 2s^2 2p^6 3s^2 3p^5$. Individual sodium and chlorine atoms would take up a specific spin-up or spin-down configuration in a magnetic field.

When molecules are formed, atoms either share or donate electrons to create closed shells or subshells that correspond to configurations of greatest stability. Sodium chloride is a three-dimensional framework structure, illustrated in Figure 6.1. Each sodium atom has transferred one of its electrons to a chlorine atom to form a Cl^- ion with electronic structure $1s^2 2s^2 2p^6 3s^2 3p^6$, which has a complete $3p$ subshell, while sodium has become a Na^+ ion, with electronic structure $1s^2 2s^2 2p^6$, which has a closed outer $2p$ shell. The structure is bound together by the attraction of neighbouring positive and negative ions to give what is known as *ionic bonding*.

Another kind of bonding, *covalent bonding*, occurs when atoms share electrons. The general basis of this process is that, by sharing, each atom in a molecule should have a complete complement of electrons in its outer shell. We illustrate this by the molecule of

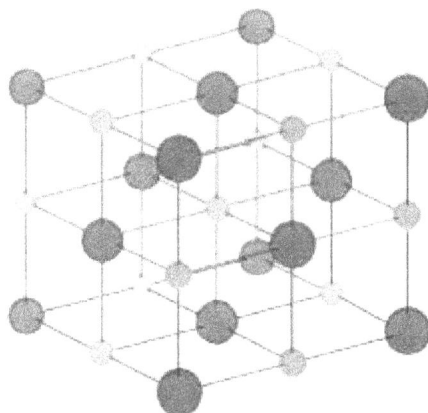

Figure 6.1 The structure of sodium chloride. Large circle = sodium, small circle = chlorine.

methane, CH_4, in which there is sharing of electrons. Carbon has an outer subshell $2p^2$ that requires the addition of four electrons to fill the $n = 2$ shell while hydrogen, with electronic structure $1s^1$, requires one extra electron to complete its outer $1s$ shell. This is done by sharing electrons as shown in Figure 6.2.

The lesson drawn from these examples is that in most cases, when molecules form, subshells are filled and, since subshells always contain even numbers of electrons, there will always be a pairing of spin-up and spin-down electrons. Hence, for the most part, molecules have no net electron-spin angular momentum and are not amenable to examination by the technique of *electron spin resonance* (ESR) *spectroscopy*, which is analogous to the use of NMR technique described in Chapter 5. An exception would be for a class of molecules containing elements such as scandium, described in Section 6.1.1, where, although it may fill up an *outer* subshell in forming a molecule, still has an unpaired electron in an *inner* shell.

Exercise 6.2 Draw a diagram, similar to Figure 6.2, to represent the electronic structure of ammonia, NH_3.

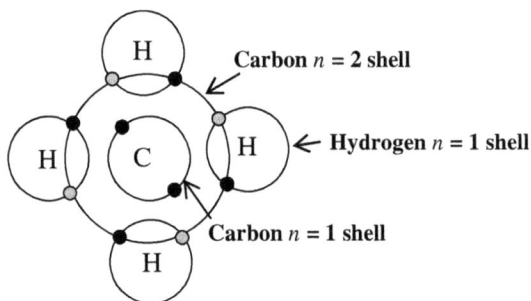

Figure 6.2 The molecule of methane held together by covalent bonds. Grey electrons are from hydrogen.

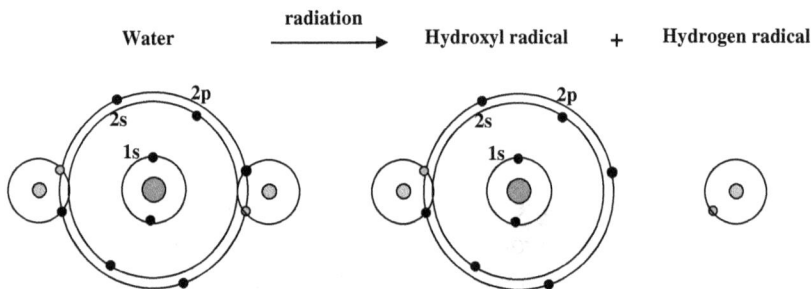

Figure 6.3 A schematic representation of the break up of water to give an OH radical plus a hydrogen atom (radical). Large grey circles are the nuclei of oxygen and hydrogen and small black circles are electrons. The size of the shells has no significance.

6.1.3 *The electronic structure of free radicals*

If energy is added in some way to a water molecule, H_2O (e.g. by exposing it to high-energy radiation), then it can be split up by

$$H_2O \rightarrow H + OH.$$

The units H and OH are free radicals, which do not have the property that each constituent atom is associated with a complete outer shell or subshell of electrons. Figure 6.3 shows a schematic representation of the electronic structure of the OH radical and it will be seen that the $2p$ subshell contains five electrons rather than the six required to fill it. That being the case, the electron content of the free radical cannot consist of pairs of electrons with balancing spins and the radical will have a net spin angular momentum and hence a net magnetic dipole moment.

Free radicals are very reactive, as they readily combine with other materials to form a stable chemical compound with complete shells. They are very common in the tails of comets (e.g. OH, CO and NH), where they form due to the disruption of stable molecules by solar radiation. They have an extended lifetime because it takes some time for them to come into contact with anything with which to react in the rarefied environment of a comet tail. In a terrestrial context they can be formed by synthesis with reagents in very dilute solution or in a rarefied gas; by break-up due to high-energy radiation or electrical discharges; or just as intermediate stages in chemical reactions in which they have a transient existence.

Exercise 6.3 Draw a diagram, similar to that of OH in Figure 6.3, to represent the electronic structure of the CH_3 radical.

6.2 The Basic Theory of Electron Spin Resonance

The theory of electron spin resonance (ESR) — also known as electron paramagnetic resonance (EPR) — closely follows that of nuclear magnetic resonance, dealt with in Chapter 5. Thus, corresponding to Equation (5.6), the magnetic moment of a free electron (one not associated with an atomic nucleus) is

$$\mu_e = -g_e s \mu_B, \tag{6.2}$$

in which the Landé g-factor for an electron, g_e, is 2.0023 and the electron-spin quantum number, s, is either $+\frac{1}{2}$ or $-\frac{1}{2}$. If the free electron is in a magnetic field of strength B, then the energy associated with each value of s, corresponding to Equation (5.7), is

$$E_s = -\mu_e B = g_e s \mu_B B. \tag{6.3}$$

The difference in energy between the higher-energy state $\left(s = +\frac{1}{2}\right)$ and the lower-energy state $\left(s = -\frac{1}{2}\right)$ is

$$\Delta E = g_e \mu_B B \tag{6.4a}$$

and if the free electron is exposed to electromagnetic radiation of frequency ν such that

$$h\nu = g_e \mu_B B, \tag{6.4b}$$

then a resonance condition is achieved and transitions can occur between the two states.

Typically specimens are exposed to a radio frequency in the range 9–10 GHz and for 10 GHz the corresponding field to give resonance is

$$B = \frac{h\nu}{g_e \mu_B} = \frac{6.626 \times 10^{-34} \times 10^{10}}{2.0023 \times 9.274 \times 10^{-24}} = 0.357 \text{ T}. \tag{6.5}$$

If there is an ensemble of free electrons then, from the Boltzmann distribution given in Equation (5.8), the ratio of those in the lower energy state to those in the higher energy state for the field in (6.5) and a temperature of 300 K is

$$\frac{P_{1/2}}{P_{-1/2}} = \frac{\exp\left(E_{1/2}/(kT)\right)}{\exp\left(E_{-1/2}/(kT)\right)} = \exp\left(\frac{\Delta E}{kT}\right) = \exp\left(\frac{h\nu}{kT}\right)$$

$$= \exp\left(\frac{6.626 \times 10^{-24}}{1.381 \times 10^{-23} \times 300}\right) = 1.0016.$$

We now anticipate a result that will be explained more fully in Section 7.2.1 by considering a system with two energy states that is exposed to radiation that has energy equal to the difference of energy between the two states. The radiation can have two effects: the first is absorption by a lower-energy state to take it to the higher energy state and the second is stimulation of a higher energy state to jump to the lower energy state. The rate of both transitions is proportional to the radiation density and the number in the initial state. What is shown in Section 7.2.1 is that the coefficient of proportionality is the same for the upwards and downwards transitions. For this reason, since there are more spins in the lower energy state, the absorption of radiation will exceed the emission.

The most common way to carry out an ESR experiment is to keep the frequency of the radiation fixed and to vary the magnetic field, although it can be done the other way round, keeping the field fixed and varying the frequency. Using the former technique, with the specimen irradiated by radiation of fixed frequency, resonance is indicated when there is a sharp rise in absorption. This is shown in Figure 6.4a, where the shape of the peak is a *Lorenzian function* given by

$$f(B) = \frac{W}{(B - B_0)^2 + W^2},\tag{6.6}$$

in which B_0 is the value of B at the peak and W is a measure of the width of the peak. Figure 6.4b shows the derivative of the absorption spectrum, which is the normal way of recording ESR spectra since, as can be seen, it better defines the value of the resonance field.

So far we have been dealing with ESR for a free electron but what we are really interested in is ESR as a means of determining atomic structure in the vicinity of an unpaired electron.

Exercise 6.4 What is the difference of energy of the spin-up and spin-down states of a free electron in a magnetic field of 0.5 T?

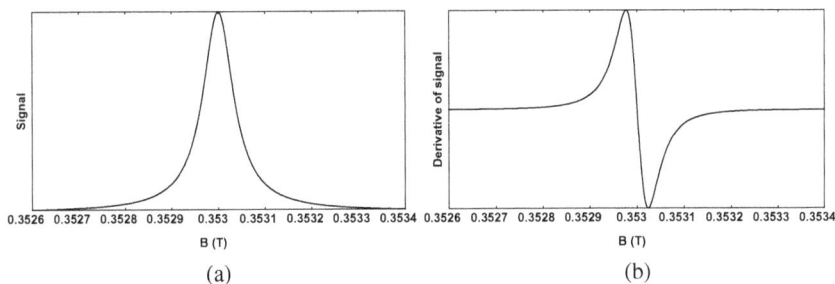

Figure 6.4 (a) An absorption spectrum for a free electron. (b) The derivative of the absorption spectrum.

Exercise 6.5   What is the ratio of spin-down to spin-up states for free electrons in a magnetic field of $0.5\,\text{T}$ at a temperature of $5\,\text{K}$?

6.3   The Form and Use of an ESR Spectrometer

There is a great deal of fairly complicated technology built into ESR spectrometers, but stripped to its basics, it is represented by Figure 6.5. The output from a microwave generator that gives a very stable frequency output such as a klystron (Section 7.3.1), travels along a waveguide to the sample. The dimensions of the waveguide and sample chamber are matched to the wavelength of the microwaves so that they contain standing waves and the sample is placed in a position where the variation in the magnetic-field component of the radiation is a maximum. Some of the radiation then travels back from the sample along the same waveguide and enters the detector. If the conditions for resonance are reached then the sample will absorb radiation and there will be a dip in the detected signal.

To record a spectrum in the form shown in Figure 6.4b the magnetic field is modulated with a small-amplitude field at a frequency

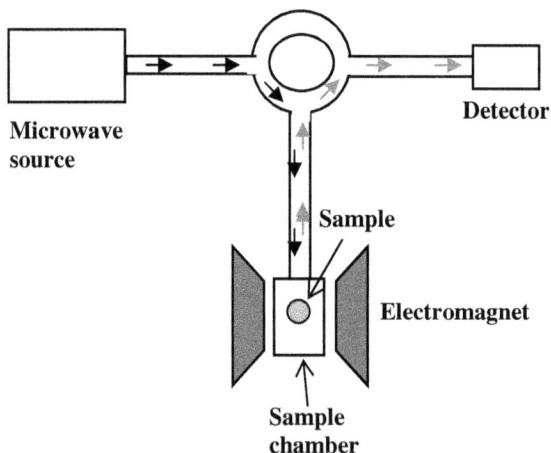

Figure 6.5 Schematic ESR equipment.

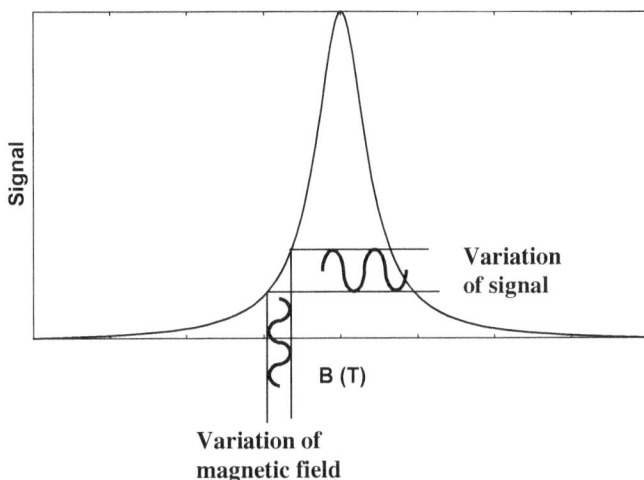

Figure 6.6 The modulation of the signal due to modulation of the magnetic field.

of about 100 kHz. If the amplitude of the modulation of the field is small compared with the width of the absorption peak, as shown in Figure 6.6, then the amplitude of the modulated signal, which is what is measured, is proportional to the slope of the absorption peak, thus emitting the signal for giving Figure 6.4b.

Exercise 6.6 Give the form of the Lorentzian function given by (6.6), find df/dB.

6.4 ESR Spectra

If all unpaired electrons behaved like free electrons then there would be nothing to be learned from ESR — all spectra would be identical and similar to that shown in Figure 6.4. However, in a radical or molecule the unpaired electrons are also affected by the local environment, in particular by nuclei with magnetic moments. There is an interaction between the electron and the neighbouring nuclei called the *hyperfine interaction*. Instead of the energy of an electron with quantum spin number s having the energy indicated by (6.2) and (6.3), due to interaction with the spin of a neighbouring nucleus it becomes

$$E_s = g_e s \mu_B B + asm, \tag{6.7}$$

where m is the nuclear-spin magnetic quantum number and a is the *hyperfine coupling constant*. Note that the additional energy is positive if s and m have the same sign and negative otherwise. It is clear that a has the dimensions of energy, although in the literature they are often given as a frequency and have to be multiplied by the Planck constant, h, to obtain the corresponding energy. The values of a usually range from about 10^{-28} to 10^{-26} J.

6.4.1 *ESR spectra for a single neighbouring nucleus with nuclear spin*

For elements that are *transition metals* and that, like scandium (Section 6.1.1), have an incomplete d subshell, there can be an unpaired electron that will interact with the nucleus of the atom of which it is part. In other cases there may be a single nucleus with no other nucleus in an equivalent position with respect to the unpaired electron. We now examine the case where the nucleus has $I = \frac{1}{2}$ so that m can only have the values $+\frac{1}{2}$ and $-\frac{1}{2}$. Energy is dependent on both s and m and we now consider the energies $E(s, m)$ where

$$E\left(+\frac{1}{2}, +\frac{1}{2}\right) = \frac{1}{2} g_e \mu_B B + \frac{1}{4} a \tag{6.8a}$$

$$E\left(+\frac{1}{2}, -\frac{1}{2}\right) = \frac{1}{2} g_e \mu_B B - \frac{1}{4} a \tag{6.8b}$$

$$E\left(-\frac{1}{2}, +\frac{1}{2}\right) = -\frac{1}{2} g_e \mu_B B - \frac{1}{4} a \tag{6.8c}$$

$$E\left(-\frac{1}{2}, -\frac{1}{2}\right) = -\frac{1}{2} g_e \mu_B B + \frac{1}{4} a. \tag{6.8d}$$

Energy E

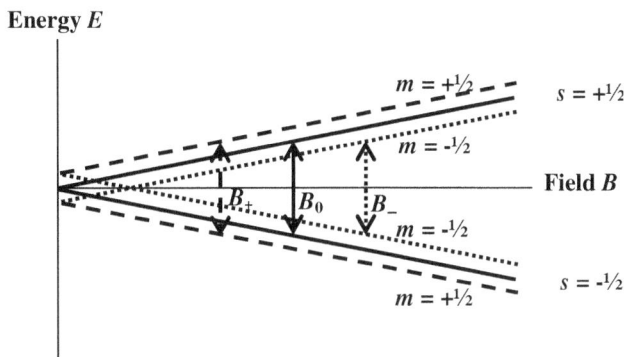

Figure 6.7 Energy levels as a function of B, with and without hyperfine interaction for $m = \frac{1}{2}$.

For a particular value of B there are four energy levels and without any constraints there would be six possible transitions, but this is not so because the transitions have to follow the following selection rules:

$$\Delta s = \pm 1$$

and

$$\Delta m = 0. \tag{6.9}$$

Therefore only two transitions are possible: between (6.8a) and (6.8c) and between (6.8b) and (6.8d). These energy levels, as functions of B, are shown in Figure 6.7.

The full line in Figure 6.7 corresponds to the energy as a function of B without hyperfine interaction. The full double-arrowed line corresponds to $h\nu$ for the imposed radiation and B_0 is the corresponding field for resonance obtained from (6.4a) as

$$B_0 = \frac{h\nu}{g_e\mu_B}. \tag{6.10}$$

The energy difference for $m = +\frac{1}{2}$ in terms of B, obtained from (6.8a) and (6.8c), is

$$\Delta E_+ = E\left(+\frac{1}{2}, +\frac{1}{2}\right) - E\left(-\frac{1}{2}, +\frac{1}{2}\right) = g_e\mu_B B + \frac{1}{2}a \tag{6.11a}$$

and similarly, from (6.8b) and (6.8d), the energy difference for $m = -\frac{1}{2}$ is

$$\Delta E_- = E\left(+\frac{1}{2}, -\frac{1}{2}\right) - E\left(-\frac{1}{2}, -\frac{1}{2}\right) = g_e \mu_B B - \frac{1}{2}a. \quad (6.11b)$$

For resonance with the applied frequency ν, giving energy difference $h\nu$, the field for $m = +\frac{1}{2}$ is

$$B_+ = \frac{h\nu - \frac{1}{2}a}{g_e \mu_B} = B_0 - \frac{a}{2g_e \mu_B} = B_0 - \frac{1}{2}Ca \quad (6.12a)$$

where $C = 1/(g_e \mu_B)$.

The resonance field for $m = -\frac{1}{2}$ is

$$B_+ = \frac{h\nu + \frac{1}{2}a}{g_e \mu_B} = B_0 + \frac{a}{2g_e \mu_B} = B_0 + \frac{1}{2}Ca. \quad (6.12b)$$

Thus the ESR spectrum shows two peaks equally spaced around B_0 separated by a difference of field Ca (Figure 6.8). For $a = 10^{-27}$ J this separation is

$$\Delta B = \frac{a}{g_e \mu_B} = \frac{10^{-27}}{2.0023 \times 9.274 \times 10^{-24}} = 5.39 \times 10^{-5}\,\text{T} = 53.9\,\mu T.$$

Exercise 6.7 An unpaired electron, irradiated with electromagnetic radiation of frequency 9 GHz, is in the vicinity of a nucleus with $I = \frac{1}{2}$. If the hyperfine coupling constant is 5×10^{-27} J then find the positions of the peaks in the ESR spectrum.

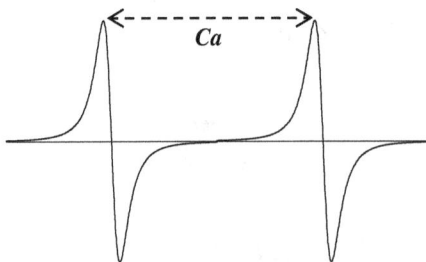

Figure 6.8 An ESR spectrum for interaction with a nucleus for which $I = \frac{1}{2}$.

6.4.2 *Many nuclei in equivalent positions*

A very common occurrence is when the unpaired electron is equidistant from two or more similar nuclei, as shown for the NH_2 radical in Figure 6.9. We now consider the electron spin interacting with the *combined spin* of the two hydrogen nuclei. These can combine as shown in Table 6.2. There will be three transitions equally spaced in energy with the $\Sigma m = 0$ transition equivalent to a transition without nuclear spin interaction and having twice the intensity of the other two.

An analysis of the situation when there are interactions with n similarly disposed spin-$\frac{1}{2}$ nuclei shows that there are $n + 1$ equally spaced resonance peaks with intensities following the pattern of a Paschal triangle, as shown in Table 6.3.

The situation is slightly more difficult to analyse if $I > \frac{1}{2}$, but it can be shown that if there are n similarly disposed nuclei, each with quantum spin number I, then the number of resonance peaks

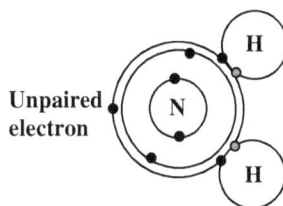

Figure 6.9 The NH_2 radical. The unpaired electron in the $2p$ subshell interacts equally with the two hydrogen nuclei.

Table 6.2 Combination of nuclear spins for two nuclei with $I = \frac{1}{2}$.

m_1	m_2	Σm
$\frac{1}{2}$	$\frac{1}{2}$	1
$\frac{1}{2}$	$-\frac{1}{2}$	0
$-\frac{1}{2}$	$\frac{1}{2}$	0
$-\frac{1}{2}$	$-\frac{1}{2}$	-1

Table 6.3 Relative intensities of ESR spectral peaks due to similar interactions with n spin-$\frac{1}{2}$ nuclei.

n	Relative intensities
1	1:1
2	1:2:1
3	1:3:3:1
4	1:4:6:4:1
5	1:5:10:10:5:1
6	1:6:15:20:15:6:1

is $2nI + 1$. In the case where $I = \frac{1}{2}$ for two nuclei, which we have already dealt with, this gives $n + 1$ lines, as previously stated.

Exercise 6.8 An unpaired electron has two similar nuclei in equivalent positions, with $I = 1$. How many ESR peaks will there be and what are their relative intensities?

6.4.3 *Two sets of non-equivalent nuclei*

A more complicated case is when there are two sets of different nuclei that are similarly disposed. As an example we take the case of the pyrazine radical anion, shown in Figure 6.10. The unpaired electron interacts with two nitrogen nuclei with $I = 1$ and a further four more distant hydrogen nuclei with $I = \frac{1}{2}$. The nitrogen nuclei alone would give five spectral lines and an analysis similar to that which followed from Table 6.3 shows that their intensities would be in the ratio 1:2:3:2:1. If alone, the more distant hydrogen nuclei with $I = \frac{1}{2}$ would also give five lines that, from Table 6.3, would have relative intensities 1:4:6:4:1.

The spectral pattern for the pyrazine radical anion, giving every combination of field displacement from nitrogen with that from hydrogen, can be built up as follows. First we take as a basic unit the five peaks from the more distant hydrogen nuclei as shown in Figure 6.11a, where we indicate each peak by a single line. The spacing of the peaks is Ca_H, where a_H is the hyperfine coupling constant

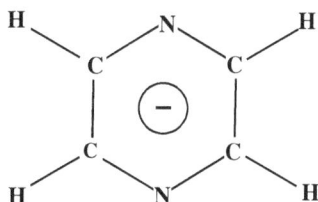

Figure 6.10 The pyrazine radical anion with the unpaired electron at its centre.

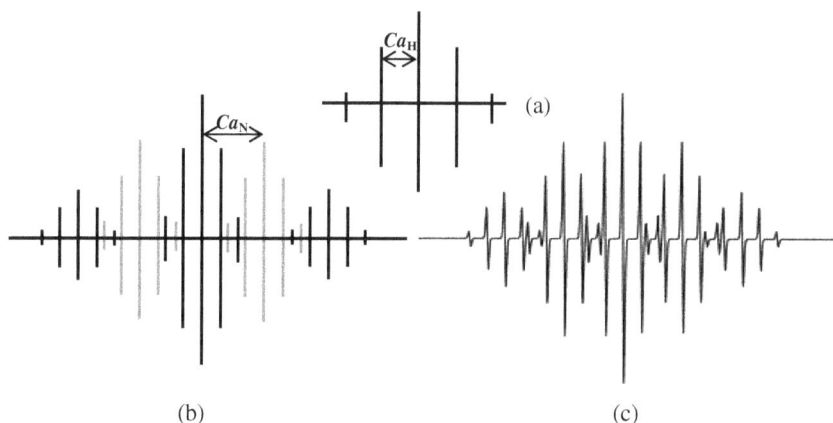

Figure 6.11 (a) A basic set of peaks from the four hydrogen nuclei. (b) A schematic simulated ESR spectrum for the pyrazine radical anion. (c) The ESR spectrum for the pyrazine radical ion.

for the hydrogen nuclei. Then, in Figure 6.11b, five of these individual units, alternately shaded, with heights in the ratio 1:2:3:2:1 are placed side by side with spacing Ca_N, corresponding to the hyperfine coupling constant for the nitrogen nuclei, a_N. The actual ESR spectrum is shown in Figure 6.11c for comparison.

Explaining how ESR spectra are formed from the features of known structures is a far simpler exercise than interpreting the spectra of unknown structures, but with experience a great deal of structural information may be obtained. The ESR technique can be used in many ways for investigations in physics, biology and chemistry, (e.g. to study radicals formed during chemical reactions). Another interesting use is in archaeology for dating teeth. When tooth enamel is exposed to radiation over long periods of time, free radicals are formed that can then be detected by ESR. From the density of free radicals in the teeth, given by the intensity of the ESR signal, fairly reliable ages can be found.

Problems 6

6.1 An unpaired electron has three similar nuclei in equivalent positions, with $I = 1$. How many ESR peaks will there be and what are their relative intensities?

6.2 The white nuclei in the molecule shown have $I = \frac{1}{2}$ and the black have $I = 1$. The hyperfine coupling constants are, respectively, in the ratio 5.5:1. Draw a diagram, similar to Figure 6.11b, showing the form of the ESR spectrum.

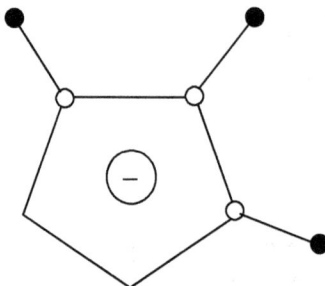

Chapter 7

Resonance with Electromagnetic Radiation

7.1 Fraunhofer Lines

If the light from the Sun is spread into a spectrum then many dark lines, known as *Fraunhofer lines*, will be seen (Figure 7.1). These lines correspond to missing wavelengths whose absences reveal the types of atoms present and something about the structure of atoms. We now describe the basic physics behind their formation.

7.1.1 *Energy levels in atoms*

The basic structure of atoms, in terms of their proton, neutron and electron content, is described in Section 5.1. The nucleus, consisting of protons and neutrons, is compact with a typical radius a few femtometres (fm; 10^{-15} m). The electrons, as represented by their state functions, are much more widely dispersed and give the atom its effective radius, of order 0.1 nanometres (nm; 10^{-9} m).

Figure 7.1 Fraunhofer lines in the solar spectrum.

The application of quantum mechanics shows that the only states that can exist are those with certain specific energies — referred to as *eigenstates*. An electron can move from one eigenstate to another but it cannot have an energy that does not correspond to an eigenstate energy. It should be noted that since electrons are bound to the nucleus their energies must be negative — positive energy would mean that they would detach themselves from the nucleus leaving behind an *ion*, an atom with one or more of its electrons missing. The more negative the electron energy is, the more tightly bound it is to the nucleus.

7.1.2 *Atomic spectra*

According to the Pauli Exclusion Principle all the electrons in an atom must have a different set of quantum numbers $(n, l, m$ and $s)$ as described in Section 6.1.1; in its lowest energy state, the *ground state*, an atom will have all the lowest possible energy states occupied. If the atoms are heated they will increase their kinetic energy, moving in random directions with a variety of speeds, with a mean energy that is proportional to the absolute temperature. As the heating is progressively increased so the atoms move faster and collisions between them become more violent. Eventually a point is reached where the energy exchanged in a collision, imparted to the highest-energy atomic electron, enables this electron to move to the next highest energy level, giving the *first excited state* (i.e. the state of the atom with the second lowest energy). The higher the temperature the greater the energy state to which the electron can be pushed. This kind of excitation of atomic electrons is illustrated in Figure 7.2.

All physical systems will, if possible, revert to the lowest possible energy state, the state of greatest stability; therefore, excited electrons will spontaneously jump back to some lower energy state, which need not be the lowest energy state. In Figure 7.2 an electron, excited from the ground state with energy E_1 to the state with energy E_4, may fall back in stages, for example from E_4 to E_3 and then from E_3 to E_1. In falling from a state with energy E_m to one with energy E_n the energy released will appear as a packet of electromagnetic

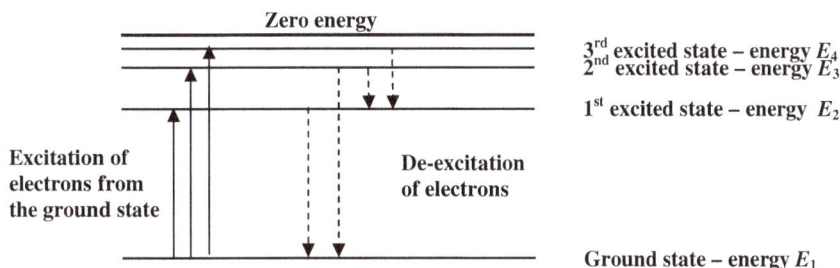

Figure 7.2 Possible excitations from the ground state (full lines) and spontaneous de-excitations (dashed lines).

Figure 7.3 A mercury spectrum.

energy, a *photon*, with frequency, ν, given by

$$h\nu = E_m - E_n, \tag{7.1}$$

where h is *Planck's constant*, 6.626×10^{-34} J s. The various frequencies emitted by a heated atom give its *atomic spectrum*, an example of which, for mercury, is shown in Figure 7.3.

Although frequency is the natural way of specifying a particular spectral line when considerations of resonance are concerned, the more common way of identifying a line is by its wavelength, connected to frequency by

$$\lambda = \frac{c}{\nu}, \tag{7.2}$$

where c is the speed of light, 2.998×10^8 m s^{-1}. The wavelengths of the pair of prominent yellow lines in Figure 7.3 are 576.98 nm and 579.07 nm.

Exercise 7.1 The ground-state energy level of hydrogen is -13.605 eV and that of the first excited state is -3.401 eV. What is the wavelength of the emitted light when an electron moves from the higher energy state to the lower? (1 eV = 1 electron volt = 1.6022×10^{-19} J.)

7.1.3 *Formation of Fraunhofer lines*

The light from the Sun comes from a very thin layer (width about
500 km) called the *photosphere*, which we see as the Sun's surface.
Light emitted below the photosphere is absorbed by material farther
out and so does not leave the Sun, while above the photosphere the
material is so diffuse that, despite the fact that it is at a very high
temperature, it emits very little light. The photons generated in the
photosphere undergo many collisions that change their energy in a
random way before they emerge, so the light emitted by the photosphere is essentially white light, containing all wavelengths. Although
the material above the photosphere is too diffuse to *emit* much light
it is a very effective *absorber* of light. For example, a sodium atom
electron in its lowest state of energy, described in the previous section
as E_1, will very readily absorb light of an energy that will *just* take it
up to energy E_2. It is a resonance effect and light of a slightly higher
or lower energy will not be absorbed. For this reason all the wavelengths that would be emitted by a heated atom will be absorbed by
the same atom if it is part of a vapour through which white light is
passing. The Fraunhofer lines tell us what types of atom exist in the
outer layers of the Sun and, since they also occur in the spectra of
distant stars, we can find out what those stars are made of.

7.2 Lasers

The term *laser* is an acronym for 'light amplification by the stimulated emission of radiation'. Lasers, which produce electromagnetic
radiation with rather special properties in the infrared to ultraviolet
region, were preceded by *masers*, which worked on the same physical principles, although with somewhat different technology, in the
microwave region — hence the 'm' in 'maser'. A laser can produce an
intense, highly collimated beam of radiation that is spatially coherent, a term that will be defined later. Here we shall first describe the
ways that an excited atom can emit radiation and then describe the
technology that produces laser emission.

7.2.1 *Spontaneous and stimulated emission*

We consider a collection of atoms in which there are N_1 atoms in a state with energy E_1 and N_2 atoms in a state with higher energy E_2. The rate of change in the number of atoms with energy E_2 due to a spontaneous jump to the lower energy state is proportional to N_2 and is given by

$$\left(\frac{dN_2}{dt}\right)_{A_{21}} = -\left(\frac{dN_1}{dt}\right)_{A_{21}} = -A_{21}N_2 \qquad (7.3)$$

where A_{21} is the *coefficient of spontaneous emission*. For spontaneous emission the emitted photons move in random directions. If a large number of emitted photons are being produced then their phases will be unrelated to one another and the emitted light will be *incoherent* — in wave terms the light emitted consists of short wave trains with no phase relationship to one another. A representation of incoherent light, travelling in one direction, is shown schematically in Figure 7.4a.

Now we assume that the atoms are being irradiated with radiation of frequency $\nu = (E_2 - E_1)/h$ with a radiation density (energy of radiation per unit volume) $\rho(\nu)$. The rate at which atoms are excited from the lower to higher energy state will be proportional to N_1 and to $\rho(\nu)$ and will be

$$\left(\frac{dN_2}{dt}\right)_{B_{12}} = -\left(\frac{dN_1}{dt}\right)_{B_{12}} = B_{12}\rho(\nu)N_1 \qquad (7.4)$$

where B_{12} is the *coefficient of absorption*. This process is similar to that which produces Fraunhofer lines.

The radiation field not only excites lower energy atoms to the higher energy but also stimulates atoms at the higher energy state to fall to the lower energy. The rate at which this occurs will be proportional to N_2 and to $\rho(\nu)$ and will be

$$\left(\frac{dN_2}{dt}\right)_{B_{21}} = -\left(\frac{dN_1}{dt}\right)_{B_{21}} = -B_{21}\rho(\nu)N_2 \qquad (7.5)$$

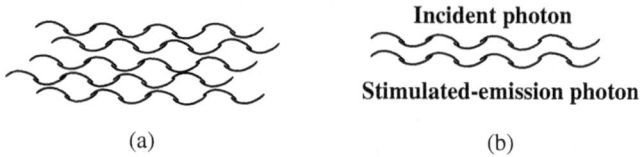

Incident photon

Stimulated-emission photon

(a)   (b)

Figure 7.4 (a) Wave trains emitted by different atoms with spontaneous emission. (b) Wave trains corresponding to an incident photon and a photon produced by stimulated emission.

where B_{21} is the *coefficient of stimulated emission*. Like the absorption that gives Fraunhofer lines, this is a resonance phenomenon since radiation with either a higher or lower frequency than ν will not cause stimulated emission to occur. An important feature of the emitted photon under stimulated emission is that it will be emitted in the same direction as the incident photon and will be *in phase* with it. This situation is represented in Figure 7.4b.

Now we consider a region which is being irradiated by radiation of density $\rho(\nu)$ and is in equilibrium so that N_1 and N_2 are constant. Then we can write

$$\left(\frac{dN_2}{dt}\right)_{A_{21}} + \left(\frac{dN_2}{dt}\right)_{B_{12}} + \left(\frac{dN_2}{dt}\right)_{B_{21}} = -A_{21}N_2 + B_{12}N_1\rho(\nu)$$
$$- B_{21}N_2\rho(\nu) = 0$$

which when rearranged gives

$$\rho(\nu) = \frac{\frac{A_{21}}{B_{21}}}{\frac{B_{12}}{B_{21}}\frac{N_1}{N_2} - 1}. \tag{7.6}$$

At this stage we call on a result from statistical mechanics — the *Boltzmann distribution* — which says that in a system of particles at absolute temperature T in equilibrium with a number of possible states, the nth one of which has energy E_n, the number with that energy is given by

$$N_n = Cg_n \exp\left(-\frac{E_n}{kT}\right) \tag{7.7}$$

where C is a constant, k is Boltzmann's constant (1.381×10^{-23} J K^{-1}) and g_n is the *degeneracy* of the energy level (i.e. the number of different states with that energy). Using this result

$$\frac{N_1}{N_2} = \frac{g_1 \exp\left(-\frac{E_1}{kT}\right)}{g_2 \exp\left(-\frac{E_2}{kT}\right)} = \frac{g_1}{g_2} \exp\left(\frac{E_2 - E_1}{kT}\right) = \frac{g_1}{g_2} \exp\left(\frac{h\nu}{kT}\right),$$

(7.8)

where ν is defined by (7.1). Substituting (7.8) into (7.6) we find

$$\rho\left(\nu\right) = \frac{\frac{A_{21}}{B_{21}}}{\frac{B_{12}}{B_{21}}\frac{g_1}{g_2} \exp\left(\frac{h\nu}{kT}\right) - 1}.$$

(7.9)

In 1900 Max Planck (1858–1947) applied quantum theory to produce the law governing the energy density inside an enclosure at absolute temperature T. In terms of frequency this is

$$\rho\left(\nu\right) = \frac{2h\nu^3/c^2}{\exp\left(\frac{h\nu}{kT}\right) - 1}$$

(7.10)

and this is the same energy density found in (7.9). Comparing the two equations it is clear that

$$\frac{A_{21}}{B_{21}} = \frac{2h\nu^3}{c^2} \quad \text{and} \quad \frac{B_{12}}{B_{21}} = \frac{g_2}{g_1}.$$

(7.11)

Although we considered the combination of spontaneous emission, absorption and stimulated emission in an equilibrium situation so that we could make a comparison with the Planck's Law Equation (7.10), the coefficients A_{21}, B_{12} and B_{21} are properties of the atom, whether or not it is in a system in equilibrium. For the working of a laser we are only interested in stimulated emission and to build it up we require the rate of stimulated emission to be greater than the rate of absorption or, from (7.4) and (7.5),

$$B_{21}\rho\left(\nu\right)N_2 > B_{12}\rho\left(\nu\right)N_1.$$

For simplicity we take $g_1 = g_2$, which gives $B_{21} = B_{12}$ and hence the condition becomes

$$N_2 > N_1,$$

(7.12)

meaning that there should be more atoms in the higher energy state than in the lower, a condition known as *population inversion*. For a laser to work this condition must be achieved, which means constantly providing energy to move atoms from a lower to a higher energy state — a process known as *pumping*.

Exercise 7.2 What is the ratio of spontaneous to stimulated emission coefficients and the ratio N_1/N_2 in a system in equilibrium when the frequency of the radiation is 6.000×10^{14} Hz and the temperature is 1,000 K? (Assume $g_1 = g_2 = 1$.)

7.2.2   *A simple laser system and uses of lasers*

The material of a laser, which provides the atoms that have the different energy levels, can be in any state: solid, liquid or gas. The way that it operates can sometimes be quite complicated and can involve more than one type of atomic species, but here we shall consider the simplest device: a gas of a single element enclosed in a tube with a mirror at each end so that light can be bounced indefinitely to-and-fro through the gas. This arrangement is shown in Figure 7.5.

The pumping of the laser is achieved by an electrical discharge through the gas, which excites atoms to the higher energy level and maintains population inversion. Initially there will just be spontaneous emission but the photons so produced will subsequently give stimulated emission from any excited atom they meet. The two mirrors are accurately adjusted so that any light that happens to move along the axis of the tube will repeatedly follow the same path to-and-fro along the axis. A photon moving in that direction may either stimulate an excited atom to produce another photon moving along

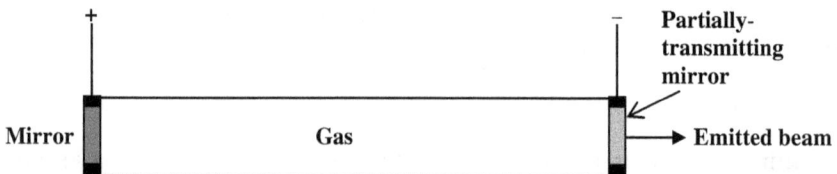

Figure 7.5 A simple gas laser.

the same path in phase with it, or be absorbed. When the pumping energy gives population inversion the stimulated emission will exceed absorption. The increasing number of coherent photons will continue to stimulate emission from other excited atoms. The mirrors increase the effective path length of the light moving along the axis, building up the intensity of the coherent light moving along the axis. Eventually the rate of stimulated emission will equal the rate at which ground state atoms are being pumped up to the higher level by the discharge, and at that point the maximum intensity of the coherent light is attained within the tube. Some part of this light passes through the partially-transmitting mirror and constitutes the output of the laser. This light is highly collimated, giving a beam that is effectively parallel over large distances, which makes it useful as a laser pointer or, more lethally, as a pointer for the targeting of weapons.

There can be many different working substances, including mixtures of materials such as helium and neon, which employ combinations of electronic energy transitions to achieve population inversion. Lasers can produce a continuous beam of radiation at comparatively low power, the so-called continuous wave (CW) mode, for a variety of uses from the readers of bar codes in supermarkets to the reading heads of DVD players. Infrared lasers using carbon dioxide (CO_2) can operate continuously at higher powers (from 30 watts to several kilowatts) for various uses, including as cutting tools in industry, and various medical applications such as surgery and dentistry.

Lasers can also be run in a pulsed mode, with the power and duration of each pulse and the repletion rate of the pulses defining the operational characteristics. The maximum power that has been achieved in a pulse is more than 1 petawatt (PW; 10^{15} W) and the pulse durations can be as short as a few femtosecond (fs: 10^{-15} s). Such lasers are used as scientific tools and have also been considered as potential weapons.

7.3 Radar

The idea of radar originated with the German inventor Christian Hulsmeyer (1881–1957) who took out a patent in 1904 for the

detection of objects by reflected radio waves. Several countries —
notably the UK, Germany, France, the USA and the Soviet Union —
took up the idea, seeking to use it for military purposes. In the period
before World War II it was suspected by the British that Germany
was trying to develop a death-ray weapon based on radio waves.
A British team led by Robert Watson-Watt (1892–1973) concluded
that, while such a device was not possible, radio waves could be
used to locate incoming aircraft. The name for the technique, RAdio
Detection And Ranging (RADAR), is now universally used and *radar*
has become a standard English word.

Radar is now used for many civil applications including:

Air traffic control	Sea traffic control
Speed traps	Intruder alarms
Navigation	Weather monitoring
Determining sea state	Locating archaeological relics
Distance and rotation of planets	Imaging planetary surfaces
Determining Earth resources	Mapping.

The military applications have expanded from those originally envis-
aged and now include:

Detection of enemy or own forces	Weapon guidance
Detection of missiles	Espionage.

We shall begin by considering the original purpose of radar systems, the detection of aircraft or naval vessels, restricting our atten-
tion to the determination of distance. Basically, the method is to
emit a radio wave towards the target and to measure the time for
it to be reflected back to its point of origin. The radio wave cannot
be continuous since there would then be no way of determining the
time interval; instead the radio wave is transmitted as a succession of
short pulses as illustrated in Figure 7.6. The duration of each pulse,
τ, is typically about 1 microsecond and the time interval between
pulses, T, is one millisecond. The number of pulses per second is
usually called the *pulse repetition frequency* (PRF) (e.g. $T = 10^{-3}$ s
gives PRF $= 10^3$ Hz). In 10^{-3} s at the speed of light (3×10^8 m s^{-1}),
a pulse covers 300 km, corresponding to a radar range of 150 km

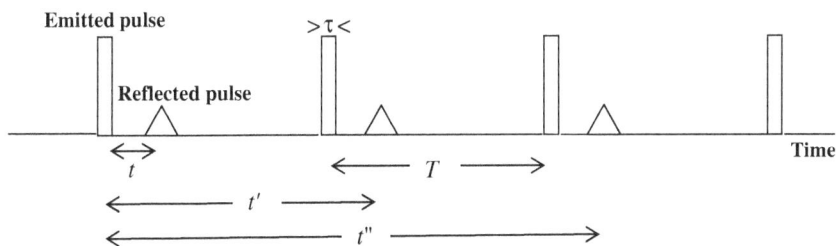

Figure 7.6 A pulsed train of radar signals with reflections.

(300 km there-and-back). For a target at a distance of 30 km the return reflection will be 200 μs behind the outward pulse. Measuring this delay time, t, gives the distance of the target.

It will be seen from Figure 7.6 that if it were not known that the target was within 150 km then there would be uncertainty about its distance; the delay in the pulse could be t', corresponding to a distance of 180 km, or t'', corresponding to a distance of 330 km. Such uncertainties are rare but can be resolved by using a number of different PRFs and looking for the common distance they indicate.

The basic process of determining the distance of a target is to generate a train of very short radio-wave pulses; to transmit the pulses to the target in a well-defined beam; and then to detect the returning radiation. Here we shall just be concerned with two different ways in which the short pulses can be generated, both of which involve resonance phenomena.

Exercise 7.3 A radar system operating with a PRF = 2,000 Hz receives a return signal from an aircraft 53 μs after each emitted radar pulse. It is known from the strength of the return signal that the aircraft is not more than 30 km away. What is its distance from the radar station?

7.3.1 *The klystron amplifier*

The klystron amplifier was invented in 1937 by the American brothers Russell Varian (1896–1959) and Sigurd Varian (1901–1961).

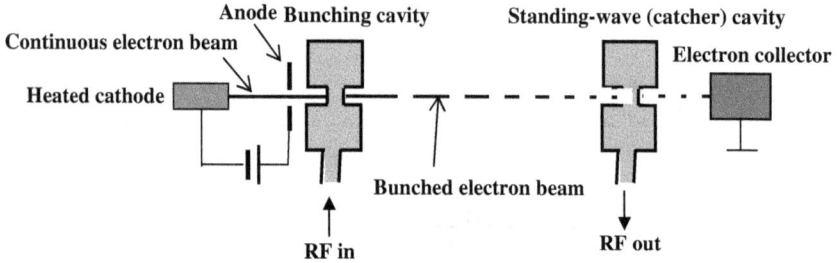

Figure 7.7 A two-cavity klystron amplifier.

The series of operations by which a klystron amplifier creates a high-power set of pulses of radio-frequency (RF) radiation can be followed by reference to the schematic representation of the device shown in Figure 7.7 and is outlined here:

(1) A beam of electrons produced by a heated cathode is accelerated through several tens of kilovolts between the cathode and anode.

(2) This beam is passed into a cavity into which is fed, through a waveguide, a fairly weak RF field that is at the resonant frequency of the cavity. Depending on the direction of the field in the cavity, when the electrons pass through they may either be accelerated or decelerated.

(3) In the space between the two cavities the accelerated electrons catch up those that have been decelerated so that as the beam traverses the space the electrons form bunches that get progressively narrower and within which the density of the electrons increases.

(4) These electrons then enter a second *catcher cavity* in which a standing-wave RF field is generated. The catcher cavity is positioned so that, when the electron bunches enter, the electric field due to the RF radiation in the cavity slows them down. The energy from the decelerated electrons generates electromagnetic radiation thereby strengthening the RF field in the catcher cavity.

(5) The catcher cavity energy so generated is channelled out via a *waveguide* to be picked up by an antenna and then reradiated as radar pulses into space in the required form.

The overall effect of the klystron amplifier is the efficient amplification, by the electrical energy used to accelerate the electrons, of the low power RF radiation fed into the bunching cavity into much higher power RF radiation being extracted from the catcher cavity.

7.3.2 *The cavity magnetron*

The cavity magnetron was developed in 1940 by the British physicists J.T. Randall (1905–1984) and H.A. Boot (1917–1983). It made an important contribution in World War II by greatly increasing the effectiveness of radar. It works on a different principle from the klystron and is capable of developing high power with high frequency in a small device, although at the expense of poorer frequency control. The higher frequency it generated enabled smaller objects to be detected and, because of its small size, it could be installed in aircraft that were hunting for submarines. A schematic representation of a cavity magnetron is shown in Figure 7.8.

The stages in the operation of a cavity magnetron are as follows:

(1) The cathode, in the form of a conducting rod coated with a material that emits electrons when heated, is subjected to a high negative potential from a direct current source, which can be either continuous or pulsed. The cathode is coaxial with the evacuated cylindrical anode, kept at ground potential, within which are situated the cylindrical resonant cavities.

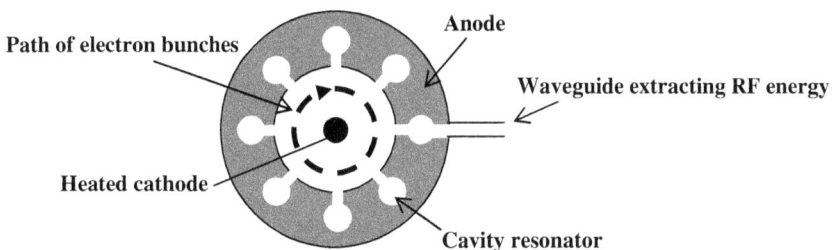

Figure 7.8 A schematic cavity magnetron.

(2) A permanent magnet creates a magnetic field parallel to the axis that, together with the radial electric field, causes the electrons to move in a circular path, as shown in the Figure 7.8.

(3) As the electrons pass the cavities in the anode they induce the production of high-frequency radiation fields within them. This is analogous to the phenomenon by which sound is produced within a bottle by blowing across its neck. The sound is at the resonant frequency of the bottle and, similarly, the frequency of the radiation field within the cavities is at their resonant frequency, which depends on their dimensions. The fields in the cavities react back on the stream of electrons, thus causing them to form bunches. The bunch structure of the electrons increases their effectiveness in producing the field within the cavities.

(4) Some of the RF energy, which is in a pulsed form, is extracted via a waveguide. For radar applications an antenna takes up this energy but the cavity magnetron is also the source of radiation in the domestic microwave cooker, in which case the radiation passes directly into the cooking chamber.

Pulsed RF radiation of between 8 and 20 cm wavelength — the so-called S-band — with peak power up to 25 MW (2.5×10^7 W) can be produced.

Exercise 7.4 A radar transmitter emits pulses of duration $0.5\,\mu s$ with an average power of $5\,\mathrm{MW}$ with $\mathrm{PRF} = 1{,}000\,\mathrm{Hz}$. What is the average power output of the transmitter?

7.4   The Anomalous Scattering of X-rays

An important scientific development in the twentieth century was that of X-ray diffraction techniques for determining the arrangement of atoms in crystals. These enable the atomic structures of proteins and even structures as large as viruses to be determined, and gave the structure of deoxyribonucleic acid (DNA), the blueprint for life. Every individual atom in the structure scatters the X-rays in all directions and, taking into account the three-dimensional periodic nature of crystals, the effect is that the total diffraction pattern from crystals consists of discrete beams of diffracted radiation. The directions of these beams indicate the size and shape of the repeat unit that forms the crystal (the *unit cell*) and the intensities of the beams contain information about the arrangement of atoms within each of the repeat units.

Deriving a crystal structure from diffraction data is not a straightforward operation and X-ray crystallographers depend on many kinds of mathematical, chemical and physical pieces of information to solve a particular structure[1]. Here we shall just be concerned with the scattering of X-rays by a single atom and how it varies with the frequency of the X-radiation. A proper treatment involves quantum mechanics, but the rather simpler classical treatment that we use here will bring out the essential features of what is termed *anomalous scattering*, one of the many important tools that can be used for structure determination.

7.4.1 *Scattering from a free electron*

We consider an electromagnetic wave, applying an alternating electric field of the form $E \exp(i\omega t)$, falling on a free electron, of mass m and charge e. The motion of the electron comes from

$$m\ddot{x} = Ee \exp(i\omega t)$$

giving the solution

$$x = -\frac{Ee}{\omega^2 m} \exp(i\omega t) = \frac{Ee}{\omega^2 m} \exp\{i(\omega t + \pi)\}, \qquad (7.13)$$

which shows that the electron oscillates π out of phase with the field. An accelerating electric charge generates electromagnetic radiation so the oscillating electron produces radiation with angular frequency ω, π out of phase with the field and scattered in all directions in space, but most strongly in the forward direction (Figure 7.9).

---

[1]M.M. Woolfson (1997) An Introduction to X-ray Crystallography, 2nd Edition, Cambridge University Press.

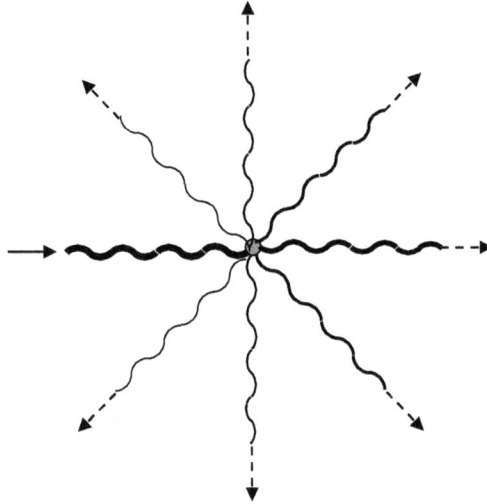

Figure 7.9 The oncoming electromagnetic wave (thickest line) is scattered by the free electron in all directions, most strongly in the forward direction and progressively weaker for higher angles of scatter.

Atomic electrons are not free but are bound to the nucleus. The strength with which an electron is bound depends on its distance from the nucleus; the farther it is from the positive charge of the nucleus the weaker is the coulomb force, and the force is further weakened by the shielding effect of the electrons further in, which acts to reduce the effective nuclear charge. Atomic electrons have a natural frequency that is related to the strength with which they are bound; the more tightly they are bound the higher is their natural frequency.

7.4.2 *Scattering from a bound electron*

Now we consider scattering from an atomic electron for which the natural angular frequency is ω_0. The equation of motion for this electron in an electric field is

$$m\ddot{x} + g\dot{x} + m\omega_0^2 x = E e \exp(i\omega t), \qquad (7.14)$$

where g is a damping constant. We seek a solution of this equation in the form $x = A \exp(i\omega t)$, which gives $\dot{x} = i\omega A \exp(i\omega t)$ and $\ddot{x} = -\omega^2 A \exp(i\omega t)$.

Substituting these values into (7.14) and rearranging gives

$$A = \frac{E_0 e}{m\left(\omega_0^2 - \omega^2\right) + ig\omega} = \frac{E_0 e\left\{m\left(\omega_0^2 - \omega^2\right) - ig\omega\right\}}{m^2\left(\omega_0^2 - \omega^3\right)^2 + g^2\omega^2}. \quad (7.15)$$

Now we express the vibration in units of the vibration from a free electron by dividing (7.15) by (7.13) to give

$$x_e = -\frac{\omega^2 m^2\left(\omega_0^2 - \omega^2\right)}{m^2\left(\omega_0^2 - \omega^2\right)^2 + g^2\omega^2} + i\frac{g\omega^3 m}{m^2\left(\omega_0^2 - \omega^2\right)^2 + g^2\omega^2}.$$

We now express this in the form

$$x_e = 1 + \varepsilon' + i\varepsilon'',$$

where ε' and ε'' are the real and imaginary parts of the additional scattering over and above that from a free electron. Thus

$$\varepsilon' = -\frac{\omega^2 m^2\left(\omega_0^2 - \omega^2\right)}{m^2\left(\omega_0^2 - \omega^2\right)^2 + g^2\omega^2} - 1 \quad (7.16a)$$

and

$$\varepsilon'' = \frac{g\omega^3 m}{m^2\left(\omega_0^2 - \omega^2\right)^2 + g^2\omega^2}. \quad (7.16b)$$

Writing $\omega_0/\omega = k$ (or in terms of wavelength, as preferred by crystallographers, $k = \lambda/\lambda_0$) we have

$$\varepsilon' = -\frac{m^2\left(k^2 - 1\right)}{m^2\left(k^2 - 1\right)^2 + g^2 k^2/\omega_0^2} - 1 \quad (7.17a)$$

and

$$\varepsilon'' = \frac{mgk/\omega_0}{m^2\left(k^2 - 1\right)^2 + g^2 k^2/\omega_0^2}. \quad (7.17b)$$

If $\omega \gg \omega_0$ (k small) and the damping factor is small then both ε' and ε'' are small and the electron behaves very like a free electron. The most tightly bound electrons in an atom are those in the K shell (principal quantum number $n = 1$), those closest to the nucleus; with the normal range of X-ray frequencies used for diffraction the other electrons in an atom will give $\omega \gg \omega_0$ (k small) and so scatter like free electrons. However, for the K electrons, in particular for higher atomic number atoms with more charge in their nuclei, it is possible to use X-rays of a frequency such that ω and ω_0 are comparable and then the scattering from the K electrons is unlike that from a free electron. To explore this numerically we need values for g, ω and ω_0.

If a thin slab of some element is traversed by X-rays then there is absorption and the intensity is attenuated. If the incident beam has intensity I and the change of intensity in passing through the slab is dI then we can write

$$\frac{dI}{I} = -\mu dx, \qquad (7.18)$$

where dx is the thickness of the slab and μ is the *linear absorption coefficient*. If the absorption is measured for a range of X-ray frequencies then the variation of μ is found to be as shown in Figure 7.10. Starting at point P, as the frequency is increased so the X-radiation becomes more energetic and penetrating and μ decreases. There is then a sudden sharp rise — an *absorption edge*. This is due to one of

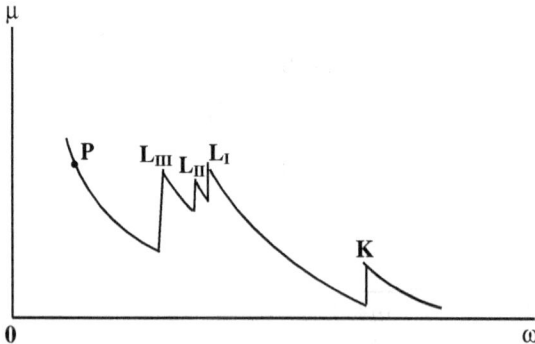

Figure 7.10 Variation of the linear absorption coefficient with frequency.

the L-shell electrons ($n = 2$) in the next layer further out than the K electrons. The sudden rise occurs when the energy of the X-ray photons is just sufficient to expel the electron from the atom. It is a similar kind of resonance absorption effect to that giving Fraunhofer lines, except that the electron is expelled from the atom rather than just being promoted to a higher energy level. Different L electrons, with different quantum numbers, have slightly different energies, giving rise to the three fairly closely spaced absorption edges. After the L_I absorption edge there is a long fall to small absorption until another absorption edge is met, this time due to the K electrons (there are two of them for all atoms except hydrogen). From a quantum-mechanical treatment of this phenomenon it turns out that the angular frequency of the absorption edge is ω_0, the natural frequency of the K electron. From the classical approach we have been using, for the resonance condition $\omega = \omega_0$ the vibration of the electron is so large it breaks free of the atom.

The absorption represents energy being turned into heat as it passes through the material and a possible source of this is the energy expended in overcoming the damping force. For the K absorption edge the K electrons will be oscillating most strongly and the work they do in opposing the damping force is the strongest influence on the fall in intensity of the X-ray beam. A general rule-of-thumb that seems to give sensible results is

$$g = 7.5 \times 10^{-15} \mu \text{ N s m}^{-1}, \qquad (7.19)$$

applied to the angular frequencies in the vicinity of the K absorption edge. It must be stressed again that a proper treatment of this topic requires the application of quantum mechanics, in which the concept of 'damping constant' does not arise, and what our classical treatment is doing is just giving a general understanding of the phenomenon.

For iron, with $\mu = 362.2 \text{ m}^{-1}$, we find $g = 2.7165 \times 10^{-12} \text{ kg s}^{-1}$. Experimentally the frequency of the X-radiation at the iron K absorption edge is 1.815×10^{18} Hz, giving $\omega_0 = 1.140 \times 10^{19} \text{ s}^{-1}$. Plots of ε' and ε'' for iron, as expressed in (7.17a) and (7.17b), are shown Figure 7.11 together with the results of a quantum-mechanical

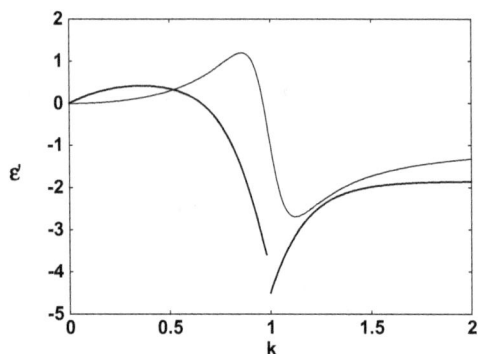

(a)

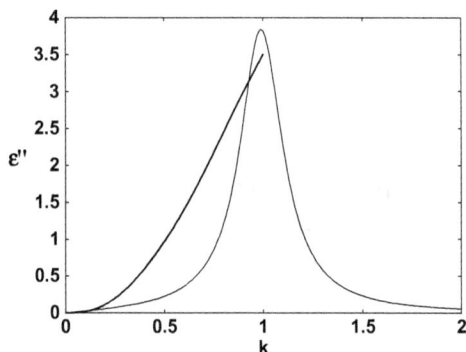

(b)

Figure 7.11 A comparison of the anomalous contributions of the K electrons of iron from the classical treatment (thin line) and a quantum-mechanical calculation (thick line) for (a) ε' and (b) ε''.

calculation. There is a resemblance in the general form of the two results with the greatest difference being that the quantum-mechanical calculation gives $\varepsilon'' = 0$ for $\omega < \omega_0$ ($k > 1$).

Anomalous scattering is an important tool in determining crystal structures. If a structure contains a few atoms of higher atomic number (e.g. selenium or iron), then by carrying out the X-ray diffraction experiment with an X-ray wavelength close to that corresponding to the K absorption edge there will be a substantial contribution from the anomalous scattering. Crystals are three-dimensional diffraction gratings and just as for an optical one-dimensional diffraction grating, where there are orders of diffraction $n = 0, \pm 1, \pm 2$ etc., so for the crystal there are diffracted beams with orders of diffraction given by three integers (called indices), h, k and l. Without anomalous scattering the intensities of the diffracted beams with indices (h, k, l) and $(-h, -k, -l)$ are equal but with anomalous scattering present the intensities are different. The differences in intensities are due to the anomalous contributions of the higher-atomic-number atoms and the differences can be used to locate their positions. Once the anomalously scattering atoms are found then there are standard procedures for progressively determining the positions of the other atoms.

Problems 7

7.1 The lowest four energy levels for hydrogen are designated as E_1 to E_4. Transitions to E_1, which is $-13.6\,\text{eV}$, from the next three highest levels give wavelengths in the ultraviolet part of the spectrum at $121.5\,\text{nm}$, $102.5\,\text{nm}$ and $97.2\,\text{nm}$. Find the energies E_2 to E_4 and show that

$$E_n = \frac{E_1}{n^2}.$$

This result comes from a quantum mechanical treatment of the hydrogen atom.

7.2 A radar transmitter emitting with $\text{PRF} = 1,000\,\text{Hz}$ records a return signal with an apparent delay of $200\,\mu\text{s}$. It immediately sends out a second burst of pulses with $\text{PRF} = 1,300\,\text{Hz}$ with an apparent delay of $123\,\mu\text{s}$. What is the distance of the target?

7.3 The linear absorption coefficient for copper at the K absorption edge is $248.5\,\text{m}^{-1}$. The K wavelength of the absorption edge for copper is $0.13808\,\text{nm}$. A diffraction experiment is carried out with wavelength $0.13750\,\text{nm}$. Determine the anomalous scattering coefficients ε_1 and ε_2. (Electron mass $= 9.109 \times 10^{-31}\,\text{kg}$.)

Chapter 8

Nuclear Physics, Radiation and Particle Physics

8.1 The Beginning of Nuclear Physics

In 1899 the New Zealand physicist Ernest Rutherford (1871–1937) discovered that the radioactive elements uranium and thorium were emitting two distinct types of radiation, which he described as α- and β-radiation. They could be distinguished by the way they behaved in a magnetic field. The β-radiation was deflected in the field and could be identified as negatively-charged electrons — hence particles rather than radiation. Later, when he used stronger fields, Rutherford found that α-radiation was also deflected and consisted of positively charged particles that we now know are helium nuclei.

Ernest Rutherford.

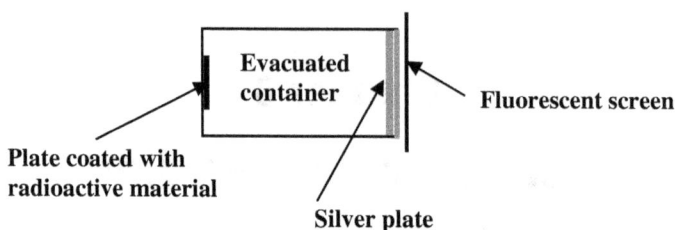

Figure 8.1 Rutherford's 1919 apparatus.

In 1919 Rutherford was carrying out an experiment to measure the range of α-particles using the apparatus shown in Figure 8.1. The α-particles were emitted by a metal plate coated with radioactive material and they travelled through an evacuated container. The range of α-particles in air was known to be 6 cm, and the silver plate at the end of the container had the same stopping power, so that α-particles should not have been able to leave the far end by passing through the silver plate. However, scintillations were observed on the fluorescent screen placed just beyond the silver plate and from their appearance it was deduced that they were produced by 'fast hydrogen atoms', which we now identify as protons. Rutherford assumed that they came from the metal plate — the source of the α-particles — and set about finding their range in various gasses. When he filled the container with either oxygen or carbon dioxide the rate of scintillations reduced, as expected. The surprise came when he filled the container with air for then the scintillation rate *increased*. Rutherford concluded that this had to be due to nitrogen, an 80% constituent of air and, as proof of this conclusion, when the container was filled with nitrogen the scintillation rate increased by 25%. We now know what was happening; a nuclear reaction was taking place that could be described as

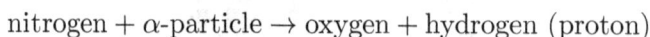

nitrogen + α-particle $\rightarrow$ oxygen + hydrogen (proton)

or, in symbolic form,

$$^{14}_{7}N + ^{4}_{2}He \rightarrow ^{17}_{8}O + ^{1}_{1}p, \tag{8.1}$$

where $^{17}_{8}O$ is a stable isotope that makes up about 0.07% of normal oxygen. The scintillations that occurred when the container was evacuated were due to the small amount of residual air that remained within it.

Rutherford was not certain what the result of his experiment had been although he knew that a transmutation had taken place; he thought that oxygen had probably been formed but he did not prove that this was so. It is strange that an experiment so simple has led, by a number of stages, from nuclear physics to particle physics, the branch of physics that operates on a huge scale, with atom-smashing machines that have dimensions of several kilometres (Section 8.6.4).

8.2 The Cockcroft–Walton Experiment

The energy of the α-particles in the Rutherford experiment was several MeV (million electron volts) and this is the order of magnitude of energy required to give the majority of nuclear reactions. In 1932 the British physicist John Cockcroft (1897–1967) and his Irish colleague, Ernest Walton (1903–1995), working at Cambridge University, designed a device, the *Cockcroft–Walton generator*, for producing high voltages by means of which protons could be accelerated through a potential difference of 710,000 V. While this energy was quite low by the usual requirements of nuclear physics, there were some nuclei that could be disintegrated by protons of this energy.

The Cockcroft–Walton experiment did not involve a resonance phenomenon but is the forerunner of many others that did, and it highlights the need for resonance to produce the high-energy particles necessary for the kinds of nuclear reactions and particle formation that are of general interest. The form of the Cockcroft–Walton apparatus is shown in Figure 8.2.

John Cockcroft (left) and Ernest Walton (right; courtesy Trinity College, Dublin).

An electric discharge ionizes hydrogen and the protons so produced pass through a hole in the plate A and are accelerated towards plate B, held at a potential 710,000 V lower than that of A. Some of these accelerated protons pass through a hole in B and strike a target of the material whose nuclei are to be disrupted. Because of the comparatively low energy of the protons, only light atoms can be used as a target; in the case of a lithium target, the reaction that occurs is

$$\,^1_1\mathrm{p} + \,^7_3\mathrm{Li} \rightarrow 2\,^4_2\mathrm{He}, \tag{8.2}$$

in which the proton combines with a lithium atom to give two helium nuclei. Cockcroft and Walton detected the presence of helium and also found that energy was produced. The products of the reaction had less mass than the original particles and the reduction of mass appeared as energy, in agreement with Einstein's mass–energy equivalence equation.

Although this was an important experiment, the limited energy of the impacting particles restricted the range of studies that could be carried out and ways were sought to produce impacting particles of much higher energy.

Exercise 8.1 With what speed, as a fraction of the speed of light, did protons strike the target in the Cockcroft–Walton experiment?

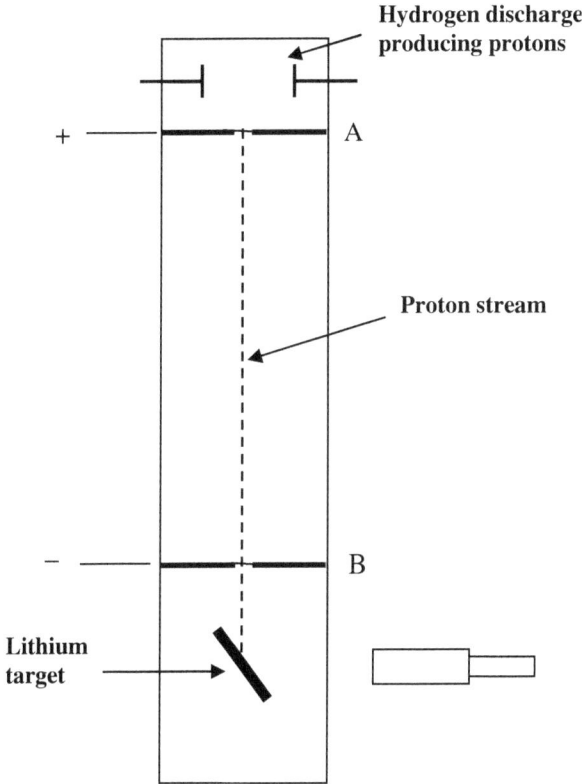

Figure 8.2 The Cockcroft–Walton apparatus.

8.3 The Cyclotron

The way that the amplitude of a child's swing was increased by impulses in resonance with its natural frequency was described in Section 2.1. The Cockcroft–Walton experiment was the equivalent of trying to reach the final amplitude by one mighty push and the restricted energy they reached was a reflection of the limitation of that approach. Modern particle accelerators employ the equivalent process of building up the amplitude of the swing by well-timed small impulses. The first such device, the *cyclotron*, was developed by the American physicist, Ernest Lawrence (1901–1958) at the University of California, Berkley, and became operational in

Ernest Lawrence.

1932 — the same year that the Cockcroft–Walton apparatus came into use.

The action of a cyclotron depends on the way that a moving charged particle is influenced by a magnetic field. In Figure 8.3a a particle with charge, q, mass, m, and velocity, v, is in the presence of a magnetic field of intensity B pointing downwards into the page. The force experienced by the particle, F, is perpendicular to both the velocity and the field vectors and has magnitude

$$F = qBv. \tag{8.3}$$

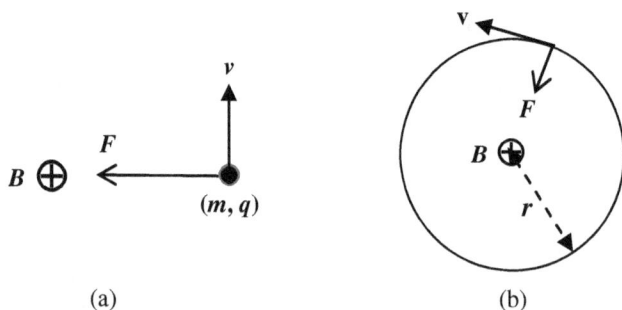

(a)　　　　　　　　　　　　　　(b)

Figure 8.3 (a) The force on a moving charged particle in a magnetic field. (b) Motion in a circle.

The effect of this force is to cause the particle to move in a circular path, as shown in Figure 8.3b. Balancing the force F with the centripetal acceleration gives

$$\frac{mv^2}{r} = qBv$$

or

$$\omega = \frac{v}{r} = \frac{qB}{m}, \tag{8.4}$$

where ω is the angular speed. The frequency of the rotational motion (i.e. the number of revolutions per unit time) is

$$f = \frac{\omega}{2\pi} = \frac{qB}{2\pi m}, \tag{8.5}$$

which is known as the *cyclotron frequency*. This frequency is independent of the radius of the orbit, but what is *not* independent of the radius is the energy of the particle, given by

$$E = \frac{1}{2}mv^2 = \frac{(qBr)^2}{2m}, \tag{8.6}$$

which is proportional to the square of the orbital radius. Equivalently, the radius of the particle's orbit is proportional to the square root of the energy.

A schematic cyclotron is shown in Figure 8.4. Charged particles are injected into a region near the centre of the cyclotron. They move in the space enclosed by two hollow D-shaped electrodes, seen in cross-section in the figure. An alternating potential is applied across the electrodes with a frequency equal to the cyclotron frequency such that, whenever a bunch of charged particles enters the gap between the electrodes it is accelerated by the electric field; when they reach the next gap the field has reversed and once again they are accelerated. The increase in energy each time the particles traverse a gap increases the radius of their orbit so they move on a quasi-spiral path. Eventually the particles emerge from the cyclotron and strike a target. The whole equipment is contained within an evacuated chamber.

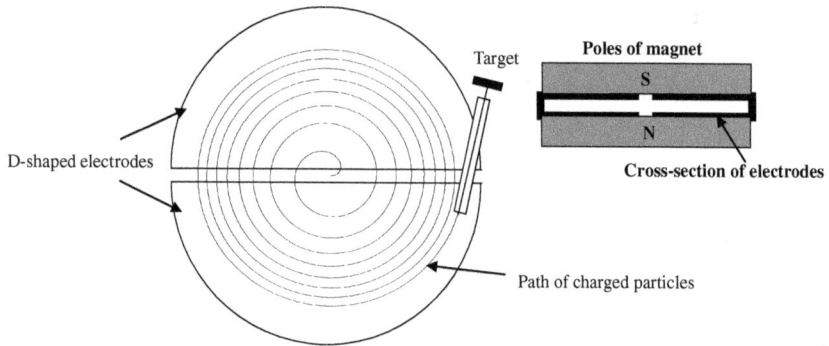

Figure 8.4 The cyclotron.

The first cyclotron built by Lawrence was a string-and-sealing-wax device with the diameter of the accelerating chamber just 12.7 cm. The energy of protons could be boosted to 80 keV with this device; it is clear from (8.6) that with such a small radius the energy of the emerging protons would be small. A succession of ever-larger cyclotrons was built by Lawrence and his co-workers over the next few years. First was a 28 cm diameter machine giving 1.2 MeV protons — the first to break the 1 MeV barrier — followed by a 68 cm machine that accelerated protons to 5 MeV and finally, in 1936, a 94 cm machine that produced protons at 8 MeV and α-particles at 16 MeV. Important new science could be done with particles of such energy: many short-lived isotopes were produced as well as the first new element, technetium $^{97}_{46}$Tc, which does not occur in nature because it has no stable isotopes. The largest modern cyclotrons are just below 20 m in diameter and can produce protons with energies up to about 350 MeV.

From the form of (8.6) it might appear that there would be an advantage in having particles with as small a mass as possible — for example, electrons. Why this is not so will now be explained.

Exercise 8.2 (i) What is the cyclotron frequency for a proton in a magnetic field of 1 T? (ii) For what radius of cyclotron accelerating chamber would this proton achieve energy 2 MeV?

8.3.1 *Maintaining resonance*

The reason for the success of the cyclotron is that the cyclotron frequency is independent of radius so that the particles arrive at the gaps between the D-electrodes just at the time that the electric field is in the right direction to accelerate them. However, there is an effect that can disturb this resonance — a consequence of Special Relativity Theory. This states that a moving particle becomes heavier and the equation governing this is

$$m = \frac{m_0}{\sqrt{1 - v^2/c^2}},\tag{8.7}$$

where m_0 is the *rest mass* of the particle, m is its mass at speed v and c is the speed of light. Rearranging we find

$$v = c\sqrt{1 - \frac{m_0^2}{m^2}}.\tag{8.8}$$

From (8.5) we can express v in terms of the cyclotron frequencies corresponding to m and m_0 (f and f_0, respectively). This gives

$$v = c\sqrt{1 - \frac{f^2}{f_0^2}}.\tag{8.9}$$

If resonance is to be maintained between the motion of the particles then, assuming a fixed frequency of the applied potential, the ratio f/f_0 must not stray too far from unity. How far that ratio can deviate from unity depends on the number of complete rotations, N, that are made by the particles within the accelerating chamber and that, in its turn, depends on the accelerating potential between the electrodes, V_d, which can be from a few thousand up to a hundred thousand volts. The final energy of the particle will be $2NV_\mathrm{d}$ (two accelerations per complete rotation) so, for a given final energy, the larger is V_d the smaller is N and the more f/f_0 can be allowed to stray from unity. The value of N can vary from one to several hundred so let us take $N = 200$ and we take it that for the last few cycles of the particles motion we can tolerate f/f_0 with value about 0.999. Then, from (8.9),

$$v = 0.045c \qquad (8.10)$$

and this is independent of the mass of the particle. An electron moving at this speed has energy about $520\,\text{eV}$ while a proton at the same speed has energy $1\,\text{MeV}$.

The speed limitation imposed by Special Relativity means that greater energies can be obtained with more massive particles. Both deuterons (^2_1D), the nuclei of deuterium, and α-particles are used in cyclotrons to increase the available final particle energies.

Exercise 8.3 It is required that the cyclotron frequency of the emerging particles from a cyclotron should be no less than 0.9999 of the value for low speed particles. What is the maximum speed allowed for the emerging particles as a fraction of the speed of light?

8.3.2 *Overcoming special relativity effects*

We have seen that the requirement to maintain the resonance between the cyclotron frequency and that of the imposed alternating potential difference limits the maximum speed of the particles to a modest fraction of the speed of light. Two ways have been found to overcome this limitation by modifying the design of the cyclotron.

The synchrocyclotron

In this modification the frequency of the electric field is changed to match the decrease in the cyclotron frequency as the mass of the particles increases. To accelerate a particle to a speed of $0.5c$ would require the final frequency, f, to be related to that at small speeds by

$$\frac{f}{f_0} = \sqrt{1 - \frac{v^2}{c^2}} = 0.866 \qquad (8.11)$$

(i.e. a 13% reduction in frequency).

The isochronous cyclotron

In this design the frequency is kept fixed but the magnet design is such that the field increases with radius to keep the ratio B/m constant; that, from (8.5), keeps the cyclotron frequency constant. A Canadian cyclotron, TRIUMF, one of the largest in the world with maximum orbit radius 7.9 m, uses this technology and produces protons with energies up to 510 MeV.

8.4 Linear Particle Accelerators

The idea of a linear accelerator (LINAC), in which the energy of particles is built up by a series of accelerations through small potential differences, actually predated the building of the Cockcroft–Walton apparatus and the first cyclotron. In a LINAC, as its name suggests, the particles are accelerated while travelling on a straight path and the original cyclotron concept was partly motivated by the idea of using magnets to bend the particle paths into circles, or spirals, so making the equipment far more compact.

The basic concept is quite simple and is illustrated in Figure 8.5. A bunch of charged particles moves along the axis of a set of separated copper-tube electrodes. Whenever it passes through a gap between the tubes the potential difference across the gap is such that it accelerates the particles. As for the cyclotron, very large particle energies can be obtained without the need for a large potential difference; the effective total accelerating potential is the sum of all the potential differences across the gaps.

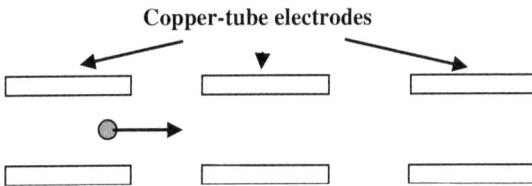

Figure 8.5 A section of a linear accelerator. The grey circle is a bunch of charged particles.

By having a very long LINAC very high energies can be obtained. The Stanford Linear Accelerator (SLAC) in California is the world's longest LINAC, with length 3.2 km, and can accelerate electrons to energy 50 GeV.[1] At such energies the speed of the particles is very high and relativistic factors come into play. The cyclotron achieves resonance at higher energies either by changing the frequency of the electrical impulses or by modifying the radial magnetic field. For the LINAC the frequency of the applied potential difference is kept constant and the resonance is maintained quite simply by variations of the lengths of the copper tubes. As the particles speed up so the tubes become longer to maintain the constant time interval between successive passages through gaps. However, once the energy builds up to relativistic levels the change of speed of the particles, and hence the changes in the lengths of the electrodes, becomes much less. The relativistic expression for kinetic energy is

$$E = mc^2 - m_0 c^2 = m_0 c^2 \left(\frac{1}{\sqrt{1 - v^2/c^2}} - 1 \right) \quad (8.12)$$

or, by rearrangement

$$\frac{v}{c} = \left\{ 1 - \left(\frac{m_0 c^2}{E + m_0 c^2} \right)^2 \right\}^{1/2}. \quad (8.13)$$

For electrons $m_0 c^2$ is about 500 keV so that for large E the right-hand side of (8.13) is close to unity and the speed of the particles varies little for large changes of E.

The charged particles in a LINAC must be in the form of compact bunches; the accelerating potential cannot be imposed on a continuous stream of particles, although it is possible to accelerate separated bunches. It was found from studies of the products of radioactive decay that there are particles with the mass of an electron but with a positive charge. These *positrons* are the *antiparticles* of electrons. If an electron and positron combine their entire mass disappears and is converted into energy. It has been found from particle

---

[1] I GeV = 10^9 eV.

physics experiments that all particles have anti-particles so that, for example, there is an antiproton and an antineutron. A LINAC can simultaneously accelerate bunches of electrons and positrons, which can be spaced along the tube so that electrons are being accelerated in a gap with an increasing potential while positrons are accelerating in a nearby gap with a decreasing potential. At the end of the tube the streams of electrons and positrons can be separated by magnetic fields and then made to collide with each other, giving annihilation and the production of γ-rays — another manifestation of matter being converted into energy.

Exercise 8.4 Electrons are injected into a LINAC with energy $1\,\mathrm{keV}$ and emerge with energy $10\,\mathrm{GeV}$. (i) What are their relative speeds at the beginning and end of their travel? (ii) If the imposed potential difference has frequency $10^9\,\mathrm{Hz}$, then what are the lengths of the copper-tube electrodes at the beginning and end of the LINAC? (Ignore the widths of the gaps between electrodes.)

8.5 Synchrotrons

The cyclotron used magnetic and electric fields to cause charged particles to move along a spiral path with ever-increasing energy while in a LINAC the particles move in a straight line. In a *synchrotron* the particles move on a closed path, which they repeatedly move around while being accelerated up to some maximum energy. A simplified illustration of a synchrotron is shown in Figure 8.6. The charged particles move around a track consisting of several straight sections linked by bending magnets that change the direction of the particles so that they go smoothly from one straight section to the next. Within the straight sections there are devices for accelerating the particles, usually by passing through microwave cavities where the particles are accelerated by moving with the microwaves in a way similar to that by which a surfer rides a wave on the sea.

To begin the process of bringing particles to very high energies, charged particles are injected into one of the straight sections from some preliminary particle accelerator (e.g. a cyclotron or linear

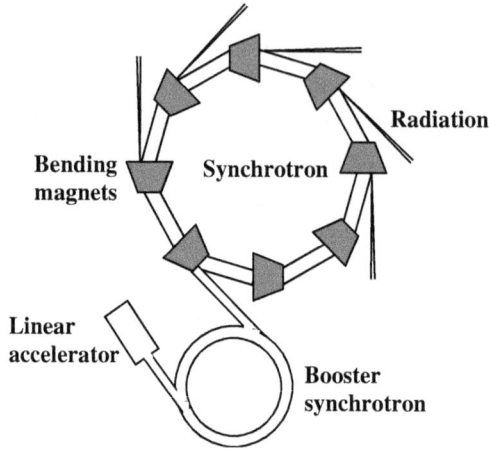

Figure 8.6 A schematic synchrotron.

accelerator) and then the energy is ramped up as they repeatedly move around the ring. The magnetic and electric fields are synchronized so that the particles keep within the track and are constantly accelerated. The limiting energy of a cyclotron is governed by its radius and the strength of the available magnetic fields. In the case of a synchrotron it is dictated by the fact that an accelerating particle radiates energy and when the bending magnets change the direction of motion of the particles they accelerate and emit electromagnetic radiation. When the rate at which energy is injected by the microwave cavities equals that emitted as radiation, then the peak particle energy is reached.

Particle physicists used the first devices of this kind, and for them the radiated energy is a nuisance and limits the energy of particles, which is their main interest. However, this radiation is intense and covers a wide range of wavelengths from the infrared to X-rays and beyond and so is useful in many other branches of science. Scientists from other fields, in particular X-ray crystallographers, became parasitic users of this *synchrotron radiation* but eventually synchrotrons were built and optimized specifically for their use. They were run as *storage rings* in which electrons are maintained in orbit for long periods of time — many hours or days — so that the emitted radiation from the bending magnets could be used for the collection of data over an extended period.

Actual storage rings usually have more than the eight bending magnets shown in Figure 8.6; the synchrotron for the Diamond Light Source at the Rutherford Appleton Laboratory in the UK, with a diameter of just over 560 m, has 48 bending magnets and 40 potential beam-lines for running experiments. The radiation coming from a storage ring has interesting and useful characteristics. It is strongly polarized, which is useful for some scientific applications, and it is not continuous but comes in bunches, separated by 10^{-8} to 10^{-9} s, which gives the possibility of carrying out time-resolved experiments on nanosecond timescales. The radiation is fanned out in a horizontal direction because radiation is emitted tangentially along the whole curved path in the bending magnets. However, the spread in the vertical direction is very small and, in radians, is

$$d\phi = m_e c^2 / E, \tag{8.14}$$

where E is the electron-beam energy, m_e the electron mass and c the speed of light. For electron-beam energies of several GeV (3 GeV for the Diamond Light Source) this spread is much less than a second of arc.

The distribution of wavelengths in the emitted radiation is always of the same general shape, as shown in Figure 8.7. The output curve is defined by a critical wavelength, λ_0, for which the total energy emitted below λ_0 equals that emitted above. This wavelength, in metres, is

$$\lambda_0 = 1.864 \times 10^{-9} / (BE^2), \tag{8.15}$$

where B is the field within the bending magnets and E is in units of GeV; the peak of the curve is at $1.4\,\lambda_0$. For Diamond, with $B = 1.4\,\mathrm{T}$, the peak is at $1.48 \times 10^{-10}\,\mathrm{m}$, which is ideal for X-ray crystallography.

The storage rings used as light sources typically use electrons with energies a few GeV because this produces a useful range of wavelengths. Various devices — called *wigglers* and *undulators* — can

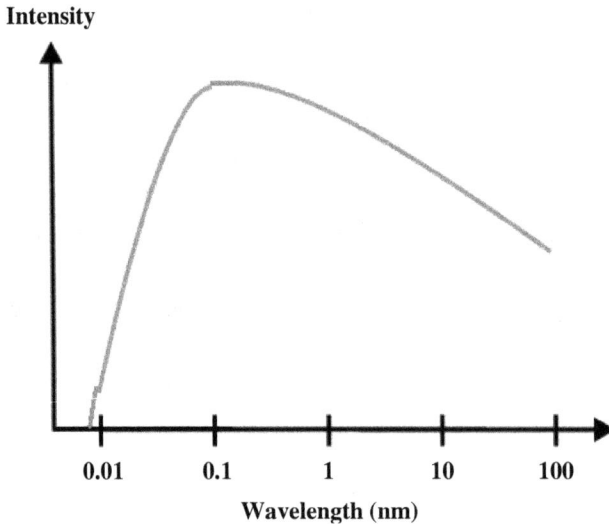

Figure 8.7 A characteristic output curve for synchrotron radiation.

be inserted in the straight sections to focus and tune the emitted radiation to particular wavelengths. However, for particle physics the requirements are for much higher energies.

Exercise 8.5 Find the horizontal and vertical spread of the radiation coming from the Diamond Light Source bending magnets.

8.6 Particles and Particle Colliders

The needs of nuclear physics and the production of synchrotron radiation can be met by particles with energies in the GeV range but for the most advanced applications in particle physics, where the goal is to break up individual nucleons (protons and neutrons) much higher energies are needed. Before discussing the structure and function of particle colliders we give here a brief summary of the field of particle physics. There are two main types of particles, *leptons* and *hadrons*, but before dealing with these, we first describe neutrinos, elusive particles with very low mass and once assumed to have zero mass.

8.6.1 *The neutrino*

Radioactive decay in which an electron is emitted can be explained by the conversion of a neutron into a proton plus electron within the nucleus. However, this simple description has several difficulties with conservation laws. Experiments in which the tracks of the residual nucleus and emitted electron were observed showed that neither energy nor momentum were being conserved and, in addition, since the neutron, proton and electron were all spin-$\frac{1}{2}$ fermions, angular momentum was also not being conserved: one spin-$\frac{1}{3}$ particle was giving rise to two spin-$\frac{1}{3}$ particles. The problem was solved by the Austrian, later American, theoretical physicist Wolfgang Pauli (1900–1958), who proposed that another fermion was emitted, a very low mass particle that was eventually called the neutrino. This proposition was later confirmed experimentally. There are also some radioactive decays in which a positron is emitted, corresponding to conversion of a proton into a neutron within the nucleus. These decay processes can be described by

$$_{0}^{1}\text{n} \rightarrow _{1}^{1}\text{p} + _{-1}^{0}\text{e} + _{0}^{0}\overline{\nu}_e \tag{8.16a}$$

and

$$_{1}^{1}\text{p} \rightarrow _{0}^{1}\text{n} + _{1}^{0}\text{e} + _{0}^{0}\nu_e. \tag{8.16b}$$

The particles $_{0}^{0}\nu_e$ and $_{0}^{0}\overline{\nu}_e$ are the *electron neutrino* and the *electron antineutrino*, respectively.

8.6.2 *Leptons*

A group of six particles (all fermions with a unit negative charge) called *leptons* are related to the electron and are fundamental in the sense that they cannot be decomposed into other particles. They are

Table 8.1 The six leptons with their masses in electron units.

1	207	3477
electron	muon	tau
electron neutrino	muon neutrino	tau neutrino

listed with their masses in Table 8.1. To each particle there is an antiparticle, so there are 12 in all.

8.6.3 *Quarks and sub-atomic particles*

Apart from protons and neutrons there are also a number of more exotic particles (e.g. *omega, lamda* and *kaon* particles) that can be decomposed into basic components. In 1964 two American theoretical physicists, Murray Gell-Mann (b. 1929) and George Zweig (b. 1937), independently proposed a model for these components, now known as *quarks*.

The six types of quark have different characteristics described as their *flavours: up* (u), *down* (d), *strange* (s), *charm* (c), *bottom* (b) and *top* (t); they are all fermions. Their charges are multiples of $\frac{1}{3}$ of an electron charge and for the six flavours they are

$$u\left(\frac{2}{3}\right) d\left(-\frac{1}{3}\right) s\left(-\frac{1}{3}\right) c\left(\frac{2}{3}\right) t\left(\frac{2}{3}\right) b\left(-\frac{1}{3}\right).$$

The corresponding *antiquarks* have opposite charges as follows

$$\bar{u}\left(-\frac{2}{3}\right) \bar{d}\left(\frac{1}{3}\right) \bar{s}\left(\frac{1}{3}\right) \bar{c}\left(-\frac{2}{3}\right) \bar{t}\left(-\frac{2}{3}\right) \bar{b}\left(\frac{1}{3}\right).$$

Particles that can be formed by combinations of quarks or antiquarks are called *hadrons*.

Combinations of three up-down quarks give protons, neutrons and their antiparticles; since quarks are fermions then combinations of odd numbers of them are also fermions. These combinations are:

Proton $\quad$ u + u + d $\quad$ has charge $\quad \dfrac{2}{3} + \dfrac{2}{3} - \dfrac{1}{3} = 1$

Antiproton $\quad$ ū + ū + d̄ $\quad$ has charge $\quad -\dfrac{2}{3} - \dfrac{2}{3} + \dfrac{1}{3} = -1$

Neutron $\quad$ d + d + u $\quad$ has charge $\quad -\dfrac{1}{3} - \dfrac{1}{3} + \dfrac{2}{3} = 0$

Antineutron $\quad$ d̄ + d̄ + ū $\quad$ has charge $\quad \dfrac{1}{3} + \dfrac{1}{3} - \dfrac{2}{3} = 0.$

Particles formed by combinations of three quarks are known as *baryons*; other baryons are *lamda particles*. The three types of lamda particle, and their compositions, are

$$\Lambda^0 \ \text{u + d + s with charge} \ \frac{2}{3} - \frac{1}{3} - \frac{1}{3} = 0$$

$$\Lambda_s^+ \ \text{u + d + c with charge} \ \frac{2}{3} - \frac{1}{3} + \frac{2}{3} = 1$$

and

$$\Lambda_b^0 \ \text{u + d + b with charge} \ \frac{2}{3} - \frac{1}{3} - \frac{1}{3} = 0$$

with corresponding antiparticles.

Quarks can also come together in pairs involving one quark and one antiquark to give various kinds of *meson*, which are *bosons* (particles with zero or integer spin). For example, *kaons*, also known as K-mesons, are formed from two quarks, one of which must be the strange quark, or its antiquark, the other being either an up or down quark or antiquark. Individual quarks have never been observed, but postulating them explains so much about particles that *are* observed that there is little doubt that they do exist.

8.6.4 *Particle colliders*

As previously explained, the use of synchrotrons for the production of radiation was an offshoot of machines designed to give very high energies that could break up normal matter to give exotic unstable particles. When used for particle physics two beams of particles can travel round the synchrotron in different directions and in non-intersecting paths. When the energy is high enough they can be deflected so as to collide and detectors of various kinds can be used to detect the resultant particles from the collision. Part of the ring of the Relativistic Heavy Ion Collider (RHIC) at the Brooklyn National Laboratory, New York, is shown in Figures 8.8 and 8.9 is an image from one of the RHIC detectors showing the tracks of thousands of particles produced by the collision of gold ions, Au^+.

The largest and most powerful particle collider is the Large Hadron Collider (LHC) built by CERN (Centre Européenne pour la Recherche Nucléaire). It has a circumference of 27 km, is part in France and part in Switzerland, and is housed in an underground tunnel of radius 3.8 m. Before particles enter the LHC they are ramped up to an energy of 450 GeV by three successive booster systems

Figure 8.8 The Relativistic Heavy Ion Collider (Brookhaven National Laboratory).

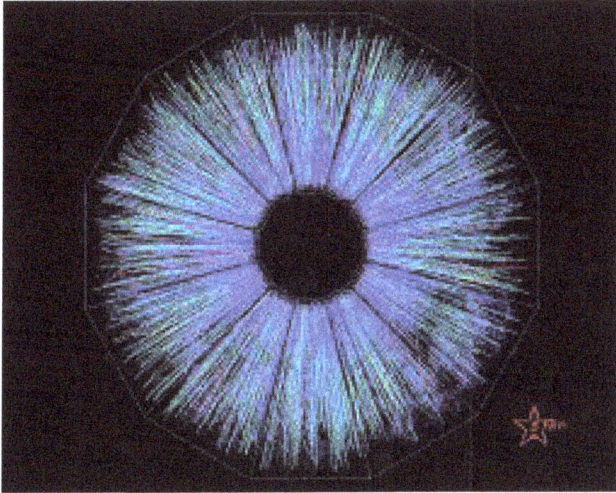

Figure 8.9 Particle tracks from the collision of very high-energy gold ions.

and within the LHC particle energies of $7\,\text{TeV}^2$ are produced, giving proton speeds of $0.999\,999\,991c$. Two beams travel in opposite directions in two pipes, both maintained under ultra-high vacuum conditions. There are 1232 bending magnets, each $15\,\text{m}$ in length, that steer the particles round the ring and 392 other *quadrupole magnets* (magnets with four poles) that focus the beams and prevent them from spreading laterally. Just before the collision of the beams takes place, other magnets concentrate the beams to increase the probability of producing collisions. All the magnets are of the superconducting kind and are cooled by liquid helium to a temperature of $1.9\,\text{K}$.

The greatest discovery made by LHC was that of the Higgs boson in 2013, a particle that bestows mass to normal matter, which was predicted by the British physicist Peter Higgs (b. 1929) some 50 years previously.

With the discovery of the Higgs boson it is not clear how much further experimental particle physics can progress. Perhaps machines of even greater power than the LHC will show new particles as yet

---

[2]I TeV $= 10^{12}$ eV.

Peter Higgs

unseen — even naked quarks — but the cost of such a machine would greatly exceed the 7.5 billion euros spent on building the LHC.

Exercise 8.6 (i) How long does it take a proton to make one complete circuit of the LHC when it is operating at full power? (ii) The field from the bending magnets is 8.4 T. What is the characteristic wavelength of the radiation they emit?

Problems 8

8.1 The accelerating chamber of a cyclotron has a radius of 5 m and the magnetic field across it has an intensity of 1.5 T. To avoid relativistic effects it is necessary for the emerging particle beam to have a speed less than $0.05c$. Facilities are available to produce singly-ionized atoms of the following elements, with their atomic masses in parentheses: calcium, Ca (40), iron, Fe (56) and copper, Cu (64).

Which ions should be used as particles to get the maximum energy and what would that energy be?

(For larger-mass nuclei take the average mass of a nucleon as the *atomic mass unit*, 1.661×10^{-27} kg.)

8.2 A uranium ion, $^{238}_{92}\text{U}^{+}$, has been accelerated to an energy of 1 TeV in the LHC.

  (i) What is its mass?

 (ii) What is its speed?

(iii) How long will it take to make a complete circuit of the LHC?

Chapter 9

The Mössbauer Effect

9.1 The Basis of the Mössbauer Effect

In Section 7.1.3 the mechanism for the formation of Fraunhofer lines in the solar spectrum was described: it is essentially a resonance mechanism whereby there is absorption of light of a wavelength corresponding to a difference in energy levels of atomic electrons. When an electron jumps from a higher to a lower energy level of some atomic species then light is emitted of a frequency given by Equation (7.1). Conversely, when light of that wavelength passes through a vapour containing that kind of atom then it is absorbed.

The atomic nucleus, consisting of protons and neutrons, also has energy levels, but the energies are much greater than those of atomic electrons and consequently the wavelengths involved are in the γ-ray region of the electromagnetic spectrum. There are emission and absorption properties, similar to those involving electron energy levels, but factors come into play that are not of importance for light emission and absorption. The approach we take here is to describe these factors in relation to the atomic electron case, with which we are familiar, and then apply them to γ-rays being emitted and absorbed by nuclei, which is the essence of the Mössbauer Effect.

9.2 Natural Line-Widths

The radiation coming from an atom when an electron falls from a higher to lower energy level is not perfectly monochromatic. The reason for this is that the transfer from the higher to lower energy state is not instantaneous; indeed it is somewhat like radioactive decay in that there is a half-life, τ, and hence an uncertainty in the time when the transition will occur. Under these circumstances the conditions of *Heisenberg's Uncertainty Principle* come into play. The most commonly expressed form of the principle relates the uncertainty of position, Δx, to that of momentum, Δp, as

$$\Delta x \Delta p \approx \hbar/2,$$

where $\hbar = h/(2\pi)$ and h is Planck's constant. Another form is

$$\Delta E \Delta t \approx \hbar/2, \tag{9.1}$$

in which ΔE is uncertainty in energy and Δt is uncertainty in time, which in this case we may replace by τ. Writing $\Delta E = h\Delta\nu$ we define the natural line-width of a emitted line as the uncertainty in frequency, $\Delta\nu_N$, given by

$$h\Delta\nu_N\tau = \hbar/2 = \frac{h}{4\pi}$$

or

$$\Delta\nu_N = \frac{1}{4\pi\tau}. \tag{9.2}$$

For transitions of electron energy levels, a typical decay time would be of order 10^{-8} s giving a line-width of order 10^7 Hz. In relation to the frequency of visible light, in the range 7.5×10^{14} and 4.3×10^{14} Hz, this is very narrow.

Exercise 9.1 The most prominent sodium spectral line (D2) is produced by the transition from the states $3p_{3/2}$ to $3s_{1/2}$. The decay time for the transition is 1.625×10^{-8} s. What is the natural width of this line?

9.3 Doppler Broadening

There is a phenomenon known as the Doppler effect, discovered by the Austrian physicist Christian Doppler (1803–1853), that is a matter of everyday experience in the world of sound. When an emergency vehicle approaches, its siren is heard at a higher pitch (frequency)

Approaching sound – wavelength shorter, pitch higher

Receding sound – wavelength longer, pitch lower

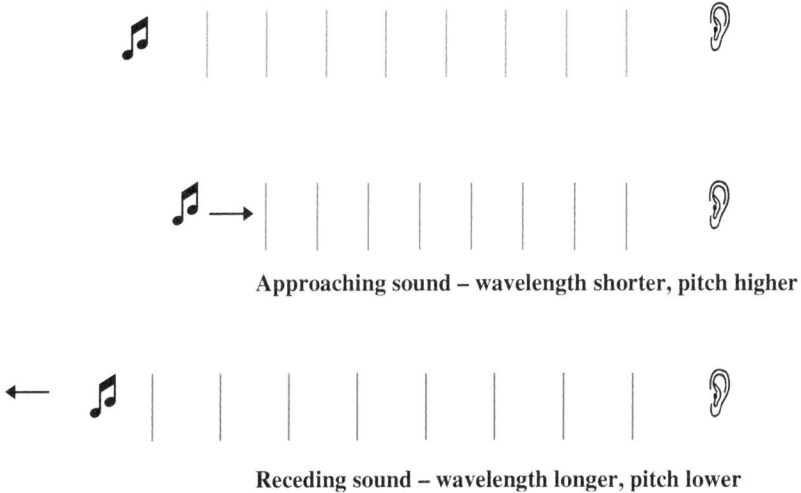

Figure 9.1 An illustration of the Doppler effect.

than when it is at rest with respect to an observer and, conversely, the pitch is lower when it moves away. In a conceptual way we may consider that the approach of a sound source compresses the sound waves (i.e. makes the wavelength smaller and hence the frequency higher) while the retreat of the source stretches them out and makes the frequency lower, as illustrated in Figure 9.1.

The Doppler effect applies to any kind of wave motion, and the relationship for electromagnetic waves between the change of frequency, $\Delta\nu$, and the radial component of velocity, V (that along the line joining source to observer) is

$$\frac{\Delta\nu}{\nu} = \frac{V}{c}, \tag{9.3}$$

where V is positive for motion towards the observer and c is the speed of light.

If light-emitting atoms or molecules are in the form of a gas then they will be moving in random directions with a distribution of speeds given by the Maxwell–Boltzmann distribution

$$P(V) = 4\pi V^2 \left(\frac{m}{2\pi kT}\right)^{3/2} \exp\left(-\frac{mV^2}{2kT}\right), \tag{9.4}$$

in which m is the mass of the particles, k is the Boltzmann constant, T the absolute temperature and the proportion of particles with speeds between V and $V + dV$ is $P(V)dV$. The root-mean-square speed of the particles is given by

$$\overline{V^2}^{1/2} = \sqrt{\frac{3kT}{m}}. \tag{9.5}$$

For each atom there will be a different direction of motion and speed but we can take the value of $2\overline{V^2}^{1/2}$ as the value to put into (9.3) to get an estimate of the Doppler broadening of emitted radiation. For example, if we take a sodium gas-discharge lamp emitting light of frequency 5.09×10^{14} Hz and with a working temperature of 1,500 K then

$$2\overline{V^2}^{1/2} = 2\sqrt{\frac{3 \times 1.381 \times 10^{-23} \times 1500}{23 \times 1.661 \times 10^{-27}}} = 2,551 \text{ m} \cdot \text{s}^{-1}$$

and inserting this value in (9.3) gives Doppler broadening as

$$\Delta\nu_D = 5.09 \times 10^{14} \times \frac{2551}{2.998 \times 10^8} = 4.33 \times 10^9 \text{ Hz}.$$

In general Doppler broadening will be greater than the natural linewidth, although cooling the source can reduce it.

Exercise 9.2 A mercury-vapour lamp, at a working temperature of 700 K, emits a prominent spectral line with wavelength 435.8 nm. What is the Doppler broadening of the line, expressed in hertz? (The mean atomic mass of mercury is 201 amu.)

9.4 The Effect of Recoil

When a stationary atom emits a photon, which has momentum $h\nu/c$, then in order to conserve momentum (which was initially zero) the atom must recoil, just as an artillery gun recoils when it fires a shell. This is illustrated in Figure 9.2.

Figure 9.2 An emitted photon and recoiling atom.

The conservation of momentum is given by

$$mV = h\nu/c \tag{9.6}$$

and conservation of energy requires the photon to lose the energy that is gained by the atom. From this the recoil change of frequency, $d\nu_R$, is given by

$$h\Delta\nu_R = -\frac{1}{2}mV^2 = -\frac{h^2\nu^2}{2mc^2}$$

or

$$\Delta\nu_R = -\frac{h\nu^2}{2mc^2}. \tag{9.7}$$

The recoil for a sodium atom emitting a photon of frequency 5.09×10^{14} Hz gives a change of frequency

$$\Delta\nu_R = \frac{6.626 \times 10^{-34} \times (5.09 \times 10^{14})^2}{2 \times 23 \times 1.661 \times 10^{-27} \times (2.998 \times 10^8)^2} = 2.50 \times 10^4 \text{ Hz},$$

which is very small and much less than both the natural line-width and Doppler broadening.

The first two of the three influences on the emitted radiation — natural line-width and Doppler broadening — affect the spread of emitted wavelengths around the mean and the third — the recoil mechanism — affects the mean frequency being emitted.

We have now established a basis for discussing the Mössbauer Effect, but before we do so we shall describe the way in which an atomic nucleus can emit a γ-ray.

Exercise 9.3 A mercury atom, with atomic mass 202 amu, emits a photon of frequency 6.879×10^{14} Hz. What is the change of frequency due to recoil?

9.5   Nuclear Emission of γ-rays

There are 44 nuclei that can emit γ-rays and we take as our example to illustrate the Mössbauer Effect the iron isotope iron-57, $^{57}_{26}$Fe, a stable isotope that forms 2.2% of natural iron. If we take a sample of iron we do not find the iron-57 component emitting γ-rays; to do this the nucleus must be in an excited state and in naturally occurring iron all nuclei will be in the ground state. It is not possible to excite a nucleus by, say, heating it; the emitted γ-rays have a frequency greater than 10 keV, which corresponds to temperatures of around 10^8 K. The way that excited iron-57 nuclei are produced is by the decay of the unstable isotope cobalt-57, $^{57}_{27}$Co. This decays by a process called *electron capture*, in which an orbital electron enters the nucleus and combines with a proton to form a neutron. This keeps the nuclear mass constant but reduces the atomic number by one, in this case giving $^{57}_{26}$Fe with the iron-57 nucleus in an excited state. The form of the transition is shown in Figure 9.3. Cobalt-57, produced in a nuclear reactor, has a half-life of 271 days. On decay by electron capture it produces an excited state of iron-57 with spin quantum number $I = 5/2$ and with energy 136 keV higher than the ground state. Of the transitions from this

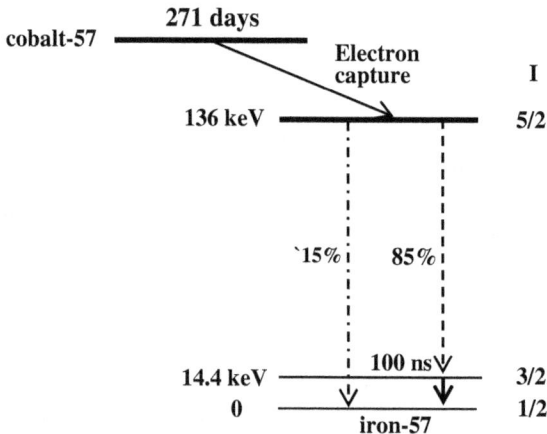

Figure 9.3 Nuclear decay of cobalt-57 to give an excited state of iron-57.

excited state, 15% are directly to the ground state with $I = 1/2$. The remaining 85% of the transitions are to the intermediate state with $I = 3/2$ and thence, with a half-life of 100 ns, to the ground state with the emission of 14.4 keV γ-rays, corresponding to a frequency $\nu_{Fe} = 3.48 \times 10^{18}$ Hz.

In our consideration of the Mössbauer Effect and how it is used as a scientific tool, we shall be taking the emission from iron-57, the one most commonly used in practice, as our example.

Exercise 9.4 The original work on the Mössbauer Effect was carried out with iridium-191, $^{191}_{77}$Ir, produced by the electron-capture decay of platinum-191 with a half-life of 2.86 days. The energy of the γ-ray emission is 129 keV. What is its frequency?

9.6 Factors Affecting γ-Ray Emission

In Sections 9.2 to 9.4 the various factors affecting the emission from atoms were dealt with, none of which had any serious effect on the use of such emissions, or the corresponding absorptions, in the infrared to ultraviolet range of the electromagnetic spectrum. We examine how these factors influence the use of the considerably higher frequencies of γ-rays as used in Mössbauer spectroscopy, bearing in mind that the energy resolution required for some applications of the technique is in the range 10^{-7} to 10^{-8} eV, corresponding to 2.4×10^6 to 2.4×10^7 Hz.

9.6.1 *Natural line-width for γ-rays*

For the iron-57 γ-ray emission the half-life for the transition from $I = 3/2$ to $I = 1/2$ is 100 ns so, from (9.2), the natural line width is

$$\Delta \nu_N = \frac{1}{4\pi \times 10^{-7}} = 7.96 \times 10^5 \text{ Hz}.$$

This corresponds to energy 3.3×10^{-9} eV and is very small in relation to the frequency, or energy, of the iron-57 γ-rays and is acceptable in terms of the required resolution.

> **Exercise 9.5** The half-life of the transition giving rise to iridium-191 γ-ray emission is 1.01×10^{-10} s. What is the natural line-width for these γ-rays?

9.6.2 *Doppler line broadening for γ-rays*

For material in the form of a gas, where individual atoms and molecules move freely, the root-mean-speed of their motion is given by (9.5). For iron at 300 K this is

$$\overline{V^2}^{1/2} = \sqrt{\frac{3 \times 1.381 \times 10^{-23} \times 300}{57 \times 1.661 \times 10^{-27}}} = 362 \, \text{m s}^{-1}.$$

Inserting $2\overline{V^2}^{1/2}$ in (9.3) to get an estimate of Doppler line broadening we find

$$\Delta \nu_D = 2 \times \frac{3.48 \times 10^{18} \times 362}{2.998 \times 10^8} = 8.4 \times 10^{12} \, \text{Hz},$$

which is far too large for the requirements of Mössbauer spectroscopy. Lowering the temperature can reduce the broadening but no amount of cooling that is attainable in practice could bring the broadening down to an acceptable level.

> **Exercise 9.6** At what temperature would iron-57 γ-ray emission have a Doppler broadening of 10^{14} Hz?

9.6.3 *The recoil frequency shift for γ-rays*

Equation (9.7) gives the reduction in the frequency of the emitted γ-ray and for iron-57 this is

$$\Delta \nu_R = \frac{6.626 \times 10^{-34} \times (3.48 \times 10^{18})^2}{2 \times 57 \times 1.661 \times 10^{-27} \times (2.998 \times 10^8)^2} = 4.71 \times 10^{11} \, \text{Hz}.$$

$$(9.8)$$

This displacement of the mean frequency of the emitted radiation will also occur in an absorption process, similar to Fraunhofer absorption, that increases the energy of the nucleus from the $I = 1/2$ state to the $I = 3/2$ state. Assuming that the frequency of the oncoming photon is precisely that required to give absorption, when it strikes the nucleus its momentum and energy change because the nucleus is set into motion. Analysis shows that the consequent reduction in its frequency is just that given by (9.8). For absorption to occur, taking into account the momentum change of the nucleus, the frequency of the arriving γ-rays would have to be $\nu_{Fe} + \Delta\nu_R$.

To summarize, if there were no Doppler broadening, so that the emitted γ-rays were almost monochromatic with only the natural line-width, then there would be no absorption since the γ-rays leaving would have frequency $\nu_{Fe} - \Delta\nu_R$ and the target iron nuclei could only absorb those with arriving frequency $\nu_{Fe} + \Delta\nu_R$.

Exercise 9.7 An iridium-191 atom emits a γ-ray photon of energy 129 keV. What is the reduction of its frequency due to recoil of the nucleus?

9.7 Mössbauer Spectroscopy

In 1958, the German physicist Rudolf Mössbauer (1929–2011) found the solution to the problems presented by Doppler broadening and recoil frequency-shifting. If both the emitting and the absorbing nuclei were bound into a lattice then these effects were reduced to such an extent that they were virtually non-existent.

Thermal energy in a lattice does not involve the independent vibration of its constituent atoms, as is the case for a gas, but cooperative vibration, like sound waves moving through the lattice. There are many different modes of vibration that can occur, associated with different energies that are constrained by quantum rules. The result of this cooperative motion is that the effective particle mass in (9.5) is not that of a single nucleus but of a vast number of nuclei, with a consequent tiny small value of $\overline{V^2}^{1/2}$ and hence virtually no Doppler broadening. The same kind of effect happens with recoil, which does not involve a single nucleus but the whole lattice so that once again the mass appearing in (9.7) is extremely large and recoil effects are virtually non-existent. With this knowledge the scene was set for the successful implementation of what is now known as Mössbauer spectroscopy.

Rudolf Mössbauer.

9.7.1 *Experimental equipment*

As has already been mentioned, a Mössbauer spectrum contains fine details and the resolution required may be of order 10^{-8} eV, corresponding to a frequency difference of order 10^6 Hz. To carry out an experiment the requirement is to be able to change γ-ray frequencies over a small range and to measure the variation of absorption in the specimen over the range. The solution is to mount the emitter on a carriage and to move it either towards or away from the absorbing sample at a fixed speed so that, due to the Doppler effect, the radiation falling on the sample changes frequency. From (9.3) to increase the frequency by 10^6 Hz for iron-57 γ-rays the emitter must move towards the sample with speed

$$V = \frac{c\Delta\nu}{\nu} = \frac{2.998 \times 10^8 \times 10^6}{3.48 \times 10^{18}} = 8.61 \times 10^{-5} \, \text{m s}^{-1},$$

or just under 0.1 mm s^{-1}.

A schematic Mössbauer spectrometer is shown in Figure 9.4.

The experiment is carried out by setting the carriage speed and then measuring the transmission through the sample. The carriage speed is repeatedly altered by small increments and in this way

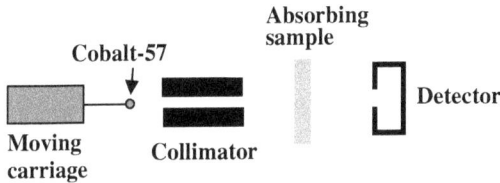

Figure 9.4 A Mössbauer spectrometer.

the maximum absorption positions are detected. It is customary in Mössbauer spectroscopy to plot absorption as a function of the carriage speed, given in the units mm s^{-1}.

9.8 Spectral Features

When the emitting and absorbing nuclei are in the same environment, the $I = 1/2$ and $I = 3/2$ spin energy levels are the same for both of them and the absorption frequency is precisely that for emission. Under such conditions, with no other influences, the absorption curve would appear as shown in the full line in Figure 9.5, with maximum absorption occurring with the carriage speed equal to zero.

However, there are various factors that influence the energies of the $I = 1/2$ and $I = 3/2$ spin states, and hence the Mössbauer spectrum, and these will now be described.

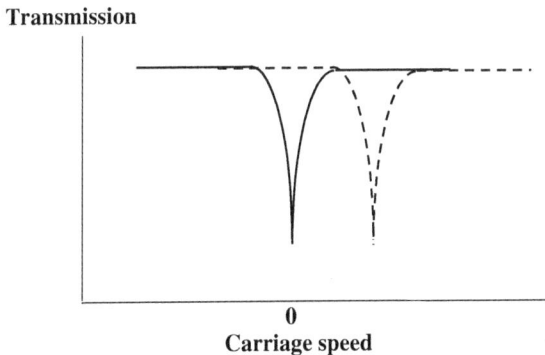

Figure 9.5 Mössbauer absorption curves for no isomer shift (full line) and isomer shift (dashed line).

9.8.1 *Isomer shift*

There is a small interaction between the nuclear charge and the electric field due to atomic electrons, in particular the $1s$ electrons that are closest to the nucleus. The $I = 1/2$ and $I = 3/2$ spin energies are affected by this interaction, but if the nuclei of the γ-ray source and the sample are in different environments within their lattices then the interactions will be different and, importantly, different for the two spin states. This is illustrated in Figure 9.6. The consequence is that the maximum-absorption frequency will differ from the emission frequency; the difference may be either positive or negative — revealed by whether the carriage speed is negative or positive for maximum absorption — giving what is called *isomer shift*. The effect of isomer shift is shown in the dashed curve in Figure 9.5.

Exercise 9.8 In a Mössbauer experiment using iron-57 γ-radiation the isomer shift corresponds to an energy of 10^{-7} eV. What must the carriage speed be to achieve maximum absorption?

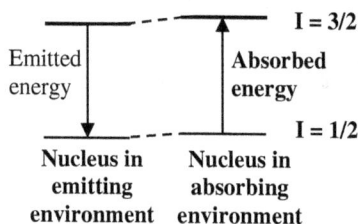

Figure 9.6 The energy differences, shown by dashed lines, associated with isomer shift are different for the two values of I.

9.8.2 *Quadrupole splitting*

A nucleus in a state with spin quantum number greater than $1/2$ has a nuclear charge distribution that is not spherically symmetric. This produces a nuclear quadrupole moment and in the presence of an asymmetric electric field, due to either the atomic electronic charge distribution or the chemical environment of the atom, this results in a splitting of the nuclear energy level. Thus the $I = 3/2$ state for iron-57 will split up into four states with $m = 3/2, 1/2, -1/2$ and $-3/2$ but the energy is only dependent on $|m|$, so the splitting is into two different energies. The energy states and transitions for iron-57 are shown in Figure 9.7a and the corresponding Mössbauer spectrum in Figure 9.7b. There are now two absorption energies but the effect of isomer shift is still present.

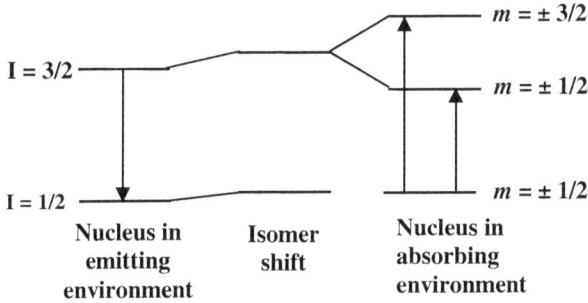

(a)

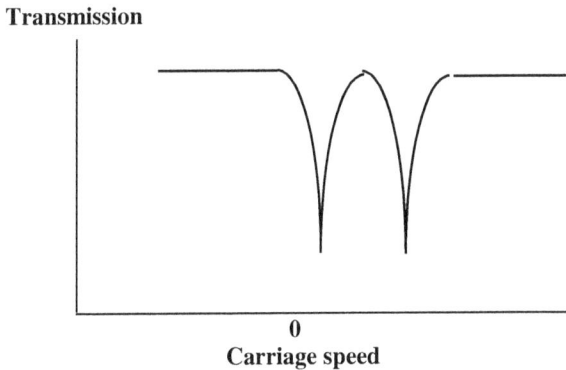

(b)

Figure 9.7 (a) Energy levels for quadrupole splitting. (b) The Mössbauer spectrum for quadrupole splitting.

9.8.3 *Magnetic splitting*

The nucleus can be subjected to a magnetic field, due either to partially filled electronic shells or to an externally applied field. In this case *Zeeman splitting* takes place and the nuclear spin can take on $2I + 1$ orientations, as shown in Figure 5.1, each of which has a different energy. Figure 9.8 shows these energy levels and the absorbing transitions that can occur between them; these transitions have to obey the selection rule that $\Delta I > 0$ and m, the magnetic quantum number, can only change by -1, 0 or 1. It will be seen that six transitions are possible and these appear as six absorption troughs in the sample-transmission plot.

Although this diagram is not an actual plot, it faithfully shows the kind of result that is possible. In particular the high resolution is evident; the distance between troughs is about 0.6 mm s^{-1}, corresponding to a frequency difference of 7×10^6 Hz, and the shapes of the individual troughs and peaks are well defined.

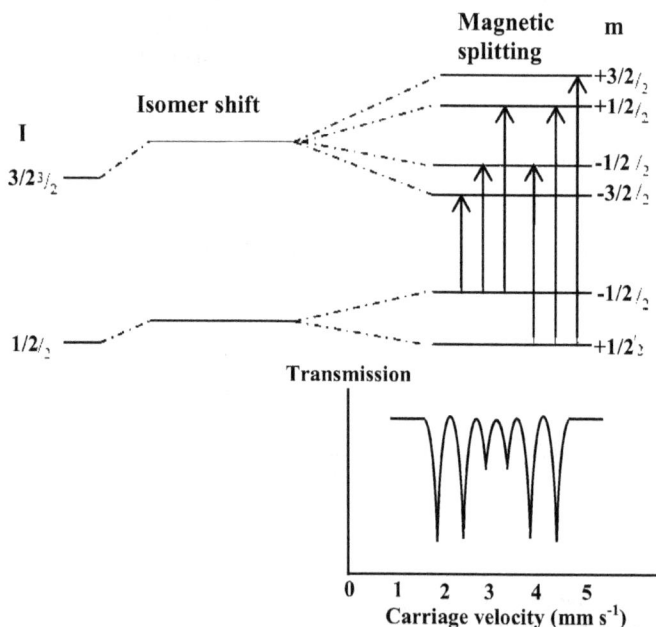

Figure 9.8 The six transitions between Zeeman-split energy levels.

9.9 Information from Mössbauer Spectroscopy

There are a limited number of isotopes that are γ-ray emitters and suitable for Mössbauer spectroscopy, including iridium-191, $^{191}_{77}$Ir, with which Rudolf Mössbauer worked when he established the technique. The resolution is remarkable, about 10^{-11} of the frequency of the γ-rays being used and, when being used by experts in the field, the technique can yield a wealth of information; in the case of iron it can reveal the valence state and many details about the environment of the iron atom. It is also possible to use the method to identify an iron mineral, although if there are two minerals with iron in the same environment then they cannot be distinguished.

An interesting application of the Mössbauer Effect was made in 1960 at Harvard University by the American physicists R.V. Pound (1919–2010) and G.A. Rebka (b. 1931) to test the *gravitational red-shift*, a consequence of Einstein's General Theory of Relativity. This gives the frequency of a photon at different distances from a gravitational centre. For small changes of distance at a large distance from the gravitational centre the red-shift, expressed in terms of frequency, $\Delta\nu$, is very closely

$$\frac{\Delta\nu}{\nu} = -\frac{\Delta\phi}{c^2}, \qquad (9.9)$$

in which $\Delta\nu$ is the change of frequency and $\Delta\phi$ the change of gravitational potential between the two positions. The form of the Harvard experiment is shown in Figure 9.9.

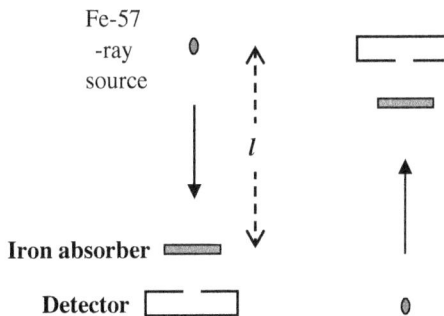

Figure 9.9 The Harvard experiment.

The experiment was run in two forms, one with the γ-ray photons moving downwards and the other with them moving upwards against the Earth's gravitational field. With the photons moving downwards the difference in gravitational potential between source and absorber is $\Delta\phi = -gl$, where g is the local acceleration due to gravity. The distance l was 22.6 m so, taking $g = 9.81\,\mathrm{m\ s^{-2}}$, the increase in frequency with the photons moving downwards was

$$\Delta\nu = \nu\frac{gl}{c^2} = 3.48 \times 10^{18}\frac{9.81 \times 22.6}{(2.998 \times 10^8)^2} = 8,584\,\mathrm{Hz}.$$

This difference of frequency was sufficient to move the absorption appreciably away from its maximum value. The emitter was mounted on a fine screw that was rotated at a constant speed, moving the source towards the absorber at a rate of a few mm per hour. The speed of the emitter was found that gave maximum absorption, which, in turn, gave a measured frequency change to be compared to the theoretical value. The experiment was also run with the photons moving upwards, when the emitter would have to move away from the absorber, and the average of the up and down experiments was compared with theory. This experiment has been repeated with refinements many times, and experiment and theory now agree to less than one part in a hundred thousand.

Exercise 9.9 In an experiment investigating the gravitational red shift γ-ray photons from Fe-57 moved at a fixed rate of 1 cm hour^{-1} towards the absorber. What is the distance between emitter and absorber to achieve the maximum absorption? ($g = 9.81\,\mathrm{m\ s^{-2}}$.)

Problem 9

9.1 In a Mössbauer experiment using iron-57 radiation and a haemoglobin crystal (iron-containing), with the sample in a magnetic field, a spectrum was produced as seen in Figure 9.8.

The velocities of maximum absorption, expressed in the units mm s^{-1}, were

$$-10.10 \quad -6.11 \quad -1.22 \quad 2.60 \quad 7.27 \quad 12.21.$$

Using Figure 9.8 as a guide, determine the energies $E(l, m)$ for the various levels shown in the figure taking $E(1/2, 1/2) = 0$. You will be able to determine $E(1/2, -1/2)$ explicitly but for the other energies you should find ΔE, where the energy is $E_0 + \Delta E$ and E_0 is the energy corresponding to velocity zero. Express energies in eV.

(Note: when independent energy estimates differ then take the average.)

Appendix I

The Binomial Theorem and Approximations

The binomial theorem is concerned with the expansion of an expression $(a + b)^n$ where n can be any real number. Since we can write

$$(a + b)^n = a^n \left(1 + a/b\right)^n,$$

there is no loss of generality if we restrict our analysis to the expression $(1 + x)^n$. The expansion can then be expressed in the form

$$(1 + x)^n = a_0 + a_1 x + a_2 x^2 + a_3 x^3 + a_4 x^4 + \cdots. \qquad \text{(A1.1)}$$

Repeatedly differentiating both sides with respect to x gives

$$n \left(1 + x\right)^{n-1} = a_1 + 2a_2 x + 3a_3 x^2 + 4a_4 x^3 + \cdots \qquad \text{(A1.2)}$$

$$n(n - 1) \left(1 + x\right)^{n-2} = 2a_2 + 3 \times 2a_3 x + 4 \times 3a_4 x^2 \ldots \qquad \text{(A1.3)}$$

$$n(n - 1)(n - 2) \left(1 + x\right)^{n-3} = 3 \times 2a_3 + 4 \times 3 \times 2a_4 x \ldots. \qquad \text{(A1.4)}$$

Putting $x = 0$ in (A1.1) to (A1.4) and equating the two sides gives

$$a_0 = 1 \quad a_1 = n \quad a_2 = \frac{n(n - 1)}{2} \quad a_3 = \frac{n(n - 1)(n - 2)}{3 \times 2}$$

and the pattern that emerges is that the coefficient a_r is

$$\frac{n(n - 1)(n - 2) \cdots (n + 1 - r)}{r!}$$

or alternatively

$$a_r = \frac{n!}{r!(n-r)!},\qquad\text{(A1.5)}$$

where $r!$ (*factorial r*) is the product of all integers from 1 to r.

In general the series has an infinite number of terms but if, and only if, n is a positive integer then after $n+1$ terms, when $r \geq n+1$ the nominator of the coefficient always has a zero factor, so the series terminates. As an example,

$$(1+x)^3 = 1 + 3x + \frac{3 \times 2}{2}x^2 + \frac{3 \times 2 \times 1}{3 \times 2}x^3 + \frac{3 \times 2 \times 1 \times 0}{4 \times 3 \times 2}x^4 + \cdots$$
$$= 1 + 3x + 3x^2 + x^3.$$

Exercise A1.1 Expand $(1+x)^6$ as a series in powers of x.

A common application of the binomial theorem is to give approximations to the value of $(1+x)^n$ under the condition that $x \ll 1$. In that case terms of order x are retained but those involving higher powers of x are not included. As a simple example we take $y = (1+x)^3$. Just taking the first two terms of the expansion of the right-hand side $y \approx 1 + 3x$. We illustrate the percentage error in using the approximation in Table A1.1

It is clear from the table that for very small values of x the approximation works well; the limit of x that can be tolerated in any particular case depends on the application being made.

Table A1.1 Approximations of $(1+x)^3$ for different values of x.

x	$(1+x)^3$	$1+3x$	% error
0.010	1.030301	1.030000	0.029
0.025	1.076891	1.075000	0.175
0.050	1.157625	1.150000	0.659
0.075	1.242297	1.225000	1.392
0.100	1.331000	1.300000	2.329

As a more complicated example we take

$$y = \frac{(3 + 2x)^4}{(1 + x)^{3/2}}$$

$$\approx 3^4(1 + 2x/3)^4(1 + x)^{-3/2}$$

$$= 3^4(1 + 4 \times 2x/3)(1 - 3x/2).$$

Now we take product of the bracketed quantities but neglect terms involving x^2. This gives

$$y \approx 3^4(1 + 8x/3 - 3x/2) = 3^4(1 + 7x/6).$$

Taking $x = 0.01$ the true value is 80.747 and the approximate value 81.945, an error of 1.48%. For $x = 0.001$ the respective values are both 81.0945 to six significant figures.

Exercise A1.2 Find an approximation for $\frac{(1+2x)^3}{(0.5+x)^{4/3}}$ in the form $a + bx$. Find the true and approximate values for $x = 0.01$ and $x = 0.005$.

Appendix II

A Program for Simulating Kirkwood Gap Formation

```
PROGRAM KIRKWOOD

C (I)    THE MAIN PROGRAMME INCLUDES THE RUNGE-KUTTA ROUTINE WITH
c        AUTOMATIC STEP CONTROL.
C (II)   SUBROUTINE "START" WHICH ENABLES INPUT OF THE INITIAL BOUNDARY
C        CONDITIONS.
C (III)  SUBROUTINE "ACC" WHICH GIVES THE ACCELERATION OF EACH BODY
C        DUE TO ITS INTERACTIONS WITH THE SUN AND JUPITER.
C (IV)   SUBROUTINE "OUT" WHICH OUTPUTS THE RESULTS TO A DATA FILE.
C
C
C THE FOUR-STEP RUNGE-KUTTA ALGORITHM IS USED. THE RESULTS OF TWO
C STEPS WITH TIMESTEP H ARE CHECKED AGAINST TAKING ONE STEP WITH
C TIMESTEP 2*H. IF THE DIFFERENCE IS WITHIN THE TOLERANCE THEN THE
C TWO STEPS, EACH OF H, ARE ACCEPTED AND THE STEPLENGTH IS DOUBLED FOR
C THE NEXT STEP. HOWEVER, IF THE TOLERANCE IS NOT SATISFIED THEN THE
C STEP IS NOT ACCEPTED AND ONE TRIES AGAIN WITH A HALVED STEPLENGTH.
C IT IS ADVISABLE, BUT NOT ESSENTIAL, TO START WITH A REASONABLE
C STEPLENGTH; THE PROGRAMME QUICKLY FINDS A SUITABLE VALUE.
C
C
      DIMENSION X(2000,3),V(2000,3),DX(2000,3,0:4),DV(2000,3,0:4),WT(4),
     +XTEMP(2,2000,3),VTEMP(2,2000,3),XT(2000,3),VT(2000,3),DELV(2000,3)

      COMMON/A/X,V,TOL,H,TOTIME,DELV,XT,VT,NB,IST,TIME,IG,XTEMP,
     +VTEMP

      DATA WT/0.0,0.5,0.5,1.0/

C SET THE INITIAL BOUNDARY CONDITION AS VALUES OF (X,Y,Z) AND (U,V,W)
C FOR EACH BODY. THIS IS DONE BY SUBROUTINE "START".
C OTHER PARAMETERS ARE ALSO SET IN "START" WHICH ALSO INDICATES
```

```
C THE SYSTEM OF UNITS BEING USED.
C
      CALL START
C
      TIME=0

C INITIALIZE ARRAYS
      DO 57 I=1,NB
      DO 57 J=1,3
      DO 57 K=0,4
         DX(I,J,K)=0
         DV(I,J,K)=0
   57 CONTINUE
C
C WE NOW TAKE TWO STEPS WITH STEP LENGTH H FOLLOWED BY ONE STEP
C WITH STEP LENGTH 2*H FROM THE SAME STARTING POINT BUT FIRST WE
C STORE THE ORGINAL SPACE AND VELOCITY COORDINATES AND TIMESTEP.
C
   25 DO 7 IT=1,2
         DO 8 J=1,NB
         DO 8 K=1,3
            XTEMP(IT,J,K)=X(J,K)
            VTEMP(IT,J,K)=V(J,K)
    8 CONTINUE

      HTEMP=H

      DO 10 NOSTEP=1,3-IT
        DO 11 I=1,4
          DO 12 J=1,NB
          DO 12 K=1,3
             XT(J,K)=XTEMP(IT,J,K)+WT(I)*DX(J,K,I-1)
   12        VT(J,K)=VTEMP(IT,J,K)+WT(I)*DV(J,K,I-1)
C
      CALL ACC
C
        DO 13 J=1,NB
        DO 13 K=1,3
          DV(J,K,I)=IT*HTEMP*DELV(J,K)
          DX(J,K,I)=IT*HTEMP*VT(J,K)

   13 CONTINUE
   11 CONTINUE
        DO 14 J=1,NB
        DO 14 K=1,3
          XTEMP(IT,J,K)=XTEMP(IT,J,K)+(DX(J,K,1)+DX(J,K,4)+2*
      +  (DX(J,K,2)+DX(J,K,3)))/6.0
          VTEMP(IT,J,K)=VTEMP(IT,J,K)+(DV(J,K,1)+DV(J,K,4)+2*
      +  (DV(J,K,2)+DV(J,K,3)))/6.0
   14 CONTINUE
```

```
   10 CONTINUE
    7 CONTINUE
C
C THE ABOVE HAS MADE TWO STEPS OF H AND, FROM THE SAME STARTING POINT,
C A SINGLE STEP OF 2*H. THE RESULTS ARE NOW COMPARED
C
      DO 20 J=1,NB
      DO 20 K=1,3
        IF(ABS(XTEMP(1,J,K)-XTEMP(2,J,K)).GT.TOL)THEN
        H=0.5*H
        GOTO 25
        ENDIF
   20 CONTINUE
C
C AT THIS STAGE THE DOUBLE STEP WITH H AGREES WITHIN TOLERANCE WITH
C THE SINGLE STEP WITH 2*H. THE TIMESTEP WILL NOW BE TRIED WITH
C TWICE THE VALUE FOR THE NEXT STEP. IF IT IS TOO BIG THEN IT WILL
C BE REDUCED AGAIN.
C
      H=2*H
      DO 80 J=1,NB
      DO 80 K=1,3
        X(J,K)=XTEMP(1,J,K)
        V(J,K)=VTEMP(1,J,K)
   80 CONTINUE

      TIME=TIME+H

      IF(INT(TIME)-INT(TIME-H).NE.0)WRITE(6,*)TIME,H

      IF(TIME.GE.TOTIME)GOTO 50
      GOTO 25
C
   50 CALL OUT
C
      STOP
      END

      SUBROUTINE START
      DIMENSION X(2000,3),V(2000,3),XTEMP(2,2000,3),VTEMP(2,2000,3),
     +DELV(2000,3),XT(2000,3),VT(2000,3)
      COMMON/A/X,V,TOL,H,TOTIME,DELV,XT,VT,NB,IST,TIME,IG,XTEMP,
     +VTEMP
C
C THE PROGRAMME AS PROVIDED IS FOR AN INVERSE-SQUARE LAW AND
C USES UNITS FOR WHICH THE UNIT MASS IS THAT OF THE SUN, THE
C THE UNIT OF DISTANCE IS THE ASTRONOMICAL UNIT (MEAN SUN-EARTH
C DISTANCE),THE UNIT OF TIME IS THE YEAR AND THE GRAVITATIONAL
C CONSTANT IS 4*PI**2.
C
```

```
      WRITE(6,'('' INPUT THE NUMBER OF BODIES'')')
      READ(5,*)NB
      WRITE(6,'('' INPUT THE INITIAL TIMESTEP [years]'')')
      READ(5,*)H
      WRITE(6,'('' INPUT TOTAL TIME FOR THE SIMULATION [years]'')')
      READ(5,*)TOTIME
C
C THE PROGRAMME ASKS THE USER TO SPECIFY A TOLERANCE, THE MAXIMUM
C ABSOLUTE ERROR THAT CAN BE TOLERATED IN ANY POSITIONAL COORDINATE
C (X, Y OR Z). IF THIS IS SET TOO LOW THEN THE PROGRAMME CAN BECOME
C VERY SLOW. FOR COMPUTATIONS INVOLVING PLANETS A TOLERANCE OF 1.0E-6
C (c. 150 KM) IS USUALLY SATISFACTORY.
C
      WRITE(6,'('' INPUT THE TOLERANCE '')')
      WRITE(6,'('' SEE COMMENT ABOVE THIS STATEMENT IN LISTING'')')
      READ(5,*)TOL
C THE FOLLOWING SETS UP THE COORDINATES OF NB BODIES UNIFORMLY
C DISTRIBUTED ON THE X-AXIS BETWEEN X = 2au AND X = 4.5au.
   TWOPI=8.0*ATAN(1.0)

   DO 31 J=1,NB
     R=2.0+FLOAT(J-1)*2.5/FLOAT(NB-1)
     X(J,1)=R
     X(J,2)=0
     X(J,3)=0
     VEL=TWOPI/SQRT(R)
      V(J,1)=0
      V(J,2)=VEL
      V(J,3)=0
   31 CONTINUE
      RETURN
      END




      SUBROUTINE ACC
      DIMENSION X(2000,3),V(2000,3),XTEMP(2,2000,3),VTEMP(2,2000,3),
     +DELV(2000,3),XT(2000,3),VT(2000,3),R(3)
      COMMON/A/X,V,TOL,H,TOTIME,DELV,XT,VT,NB,IST,TIME,IG,XTEMP,
     +VTEMP
C
C THE SUBROUTINE AS PROVIDED IS FOR AN INVERSE-SQUARE LAW AND USES
C UNITS AS DEFINED IN SUBROUTINE "START".
C
C SET THE VALUE OF G IN ASTRONOMICAL UNITS
      PI=4.0*ATAN(1.0)

      G=4*PI*PI
```

```
C SET THE POSITION OF JUPITER WITH ORBITAL RADIUS 5.20au
      XJUP=5.2*COS(2*PI/5.2**1.5*TIME)
      YJUP=5.2*SIN(2*PI/5.2**1.5*TIME)

      DO 1 J=1,NB
      DO 1 K=1,3
        DELV(J,K)=0
   1  CONTINUE
C THE FOLLOWING FINDS INTERACTIONS WITH THE SUN AND JUPITER FOR ALL
C BODIES

      DO 2 J=1,NB
        R(1)=XT(J,1)-XJUP
        R(2)=XT(J,2)-YJUP
        R(3)=0
          RRR=(R(1)**2+R(2)**2+R(3)**2)**1.5
          RSUN=(XT(J,1)**2+XT(J,2)**2+XT(J,3)**2)**1.5
      DO 4 K=1,3
C THE NEXT STATEMENT GIVES THE CONTRIBUTIONS TO THE THREE COMPONENTS
C OF ACCELERATION ON EACH BODY DUE TO THE SUN AND JUPITER
        DELV(J,K)=DELV(J,K)-G*0.001*R(K)/RRR-G*XT(J,K)/RSUN

   4  CONTINUE
   2  CONTINUE
      RETURN
      END



      SUBROUTINE OUT

      DIMENSION X(2000,3),V(2000,3),XTEMP(2,20000,3),VTEMP(2,2000,3),
     +DELV(2000,3),XT(2000,3),VT(2000,3),SMA(65)

      COMMON/A/X,V,TOL,H,TOTIME,DELV,XT,VT,NB,IST,TIME,IG,XTEMP,
     +VTEMP

      OPEN(UNIT=50,FILE='KIRKWOOD.DAT')
      DO 3 I=1,65
      SMA(I)=0
   3  CONTINUE

      PI=4.0*ATAN(1.0)
      G=4*PI**2

C CALCULATE SEMI-MAJOR AXES OF ASTEROID ORBITS AND FIND NUMBERS IN
C SMA RANGES OF 0.05au FOR SMAs BETWEEN 1.6au AND 4.8au.

      DO 1 I=1,NB
        DIS=SQRT(X(I,1)**2+X(I,2)**2+X(I,3)**2)
```

```
      ENERGY=0.5*(V(I,1)**2+V(I,2)**2+V(I,3)**2)-G/DIS
      A=-G/2/ENERGY
      IF(A.LT.1.6.OR.A.GT.4.8)GOTO 1
      N=INT((A-1.6)/0.05)+1
      SMA(N)=SMA(N)+1
    1 CONTINUE

      DO 2 J=1,65
        A=1.6+(J-1)*0.05
        WRITE(50,*)A,SMA(J)
    2 CONTINUE

      RETURN
      END
```

Appendix III

A Program for Finding the Orbits of Trojan Asteroids

```
C PROGRAMME TROJANS
        PROGRAM NBODY
C THIS IS A GENERAL N-BODY PROGRAMME WHERE INTER-BODY FORCES BETWEEN
C BODIES i AND j ARE OF THE FORM CM(i)*CM(j)*F(Rij,Vij) WHERE F IS A
C FUNCTION OF Rij, THE DISTANCE BETWEEN THE BODIES, AND Vij, THE
C RELATIVE VELOCITIES OF THE TWO BODIES.
C THE STRUCTURE OF THE PROGRAMME IS;
C
C (I)    THE MAIN PROGRAMME "NBODY"IS WHICH INCLUDES THE RUNGE-KUTTA
C        ROUTINE WITH AUTOMATIC STEP CONTROL.
C (II)   SUBROUTINE "START" WHICH ENABLES INPUT OF THE INITIAL BOUNDARY
C        CONDITIONS.
C (III)  SUBROUTINE "ACC" WHICH GIVES THE ACCELERATION OF EACH BODY
C        DUE TO ITS INTERACTIONS WITH ALL OTHER BODIES.
C (IV)   SUBROUTINE "STORE" WHICH STORES INTERMEDIATE COORDINATES AND
C        VELOCITY COMPONENTS AS THE COMPUTATION PROGRESSES.
C (V)    SUBROUTINE "OUT" WHICH OUTPUTS THE RESULTS TO DATA FILES.
C
C BY CHANGING THE SUBROUTINES DIFFERENT PROBLEMS MAY BE SOLVED.
C THE CM'S CAN BE MASSES OR CHARGES OR BE MADE EQUAL TO UNITY WHILE
C THE FORCE LAW CAN BE INVERSE-SQUARE OR ANYTHING ELSE -
C e.g. LENNARD-JONES. SEE COMMENT AT THE BEGINNING OF SUBROUTINE
C "ACC" FOR THE TYPES OF FORCES OPERATING.
C
C THE FOUR-STEP RUNGE-KUTTA ALGORITHM IS USED. THE RESULTS OF TWO
C STEPS WITH TIMESTEP H ARE CHECKED AGAINST TAKING ONE STEP WITH
C TIMESTEP 2*H. IF THE DIFFERENCE IS WITHIN THE TOLERANCE THEN THE
C TWO STEPS, EACH OF H, ARE ACCEPTED AND THE STEPLENGTH IS DOUBLED FOR
C THE NEXT STEP. HOWEVER, IF THE TOLERANCE IS NOT SATISFIED THEN THE
C STEP IS NOT ACCEPTED AND ONE TRIES AGAIN WITH A HALVED STEPLENGTH.
C IT IS ADVISABLE, BUT NOT ESSENTIAL, TO START WITH A REASONABLE
```

```
C STEPLENGTH; THE PROGRAMME QUICKLY FINDS A SUITABLE VALUE.
C
C AS PROVIDED THE PROGRAMME HANDLES UP TO 20 BODIES BUT THIS CAN BE
C CHANGED FROM 20 TO WHATEVER IS REQUIRED IN THE DIMENSION STATEMENT.
C
      DIMENSION CM(20),X(20,3),V(20,3),DX(20,3,0:4),DV(20,3,0:4),WT(4),
     +XTEMP(2,20,3),VTEMP(2,20,3),XT(20,3),VT(20,3),DELV(20,3)
      COMMON/A/X,V,TOL,H,TOTIME,DELV,XT,VT,NB,IST,TIME,IG,XTEMP,
     +VTEMP,CM
      DATA WT/0.0,0.5,0.5,1.0/
      IST=0
      OPEN(UNIT=9,FILE='LPT1')
C
C SETTING THE INITIAL BOUNDARY CONDITION CAN BE DONE EITHER
C EXPLICITLY AS VALUES OF (X,Y,Z) AND (U,V,W) FOR EACH BODY OR
C CAN BE COMPUTED. THIS IS CONTROLLED BY SUBROUTINE "START".
C OTHER PARAMETERS ARE ALSO SET IN "START" WHICH ALSO INDICATES
C THE SYSTEM OF UNITS BEING USED.
C
      CALL START
C
      TIME=0
C INITIALIZE ARRAYS
      DO 57 I=1,20
      DO 57 J=1,3
      DO 57 K=0,4
         DX(I,J,K)=0
         DV(I,J,K)=0
   57 CONTINUE
C
C WE NOW TAKE TWO STEPS WITH STEP LENGTH H FOLLOWED BY ONE STEP
C WITH STEP LENGTH 2*H FROM THE SAME STARTING POINT BUT FIRST WE
C STORE THE ORGINAL SPACE AND VELOCITY COORDINATES AND TIMESTEP.
C
   25 DO 7 IT=1,2
      DO 8 J=1,NB
      DO 8 K=1,3
         XTEMP(IT,J,K)=X(J,K)
         VTEMP(IT,J,K)=V(J,K)
    8 CONTINUE
      HTEMP=H
      DO 10 NOSTEP=1,3-IT
      DO 11 I=1,4
      DO 12 J=1,NB
      DO 12 K=1,3
         XT(J,K)=XTEMP(IT,J,K)+WT(I)*DX(J,K,I-1)
   12    VT(J,K)=VTEMP(IT,J,K)+WT(I)*DV(J,K,I-1)
C
      CALL ACC
C
```

```
      DO 13 J=1,NB
      DO 13 K=1,3
      DV(J,K,I)=IT*HTEMP*DELV(J,K)
      DX(J,K,I)=IT*HTEMP*VT(J,K)
   13 CONTINUE
   11 CONTINUE
      DO 14 J=1,NB
      DO 14 K=1,3
         XTEMP(IT,J,K)=XTEMP(IT,J,K)+(DX(J,K,1)+DX(J,K,4)+2*
     +   (DX(J,K,2)+DX(J,K,3)))/6.0
         VTEMP(IT,J,K)=VTEMP(IT,J,K)+(DV(J,K,1)+DV(J,K,4)+2*
     +   (DV(J,K,2)+DV(J,K,3)))/6.0
   14 CONTINUE
   10 CONTINUE
    7 CONTINUE
C
C THE ABOVE HAS MADE TWO STEPS OF H AND, FROM THE SAME STARTING POINT,
C A SINGLE STEP OF 2*H. THE RESULTS ARE NOW COMPARED
C
      DO 20 J=1,NB
      DO 20 K=1,3
         IF(ABS(XTEMP(1,J,K)-XTEMP(2,J,K)).GT.TOL)THEN
           H=0.5*H
           GOTO 25
         ENDIF
   20 CONTINUE
C
C AT THIS STAGE THE DOUBLE STEP WITH H AGREES WITHIN TOLERANCE WITH
C THE SINGLE STEP WITH 2*H. THE TIMESTEP WILL NOW BE TRIED WITH
C TWICE THE VALUE FOR THE NEXT STEP. IF IT IS TOO BIG THEN IT WILL
C BE REDUCED AGAIN.
C
      H=2*H
      DO 80 J=1,NB
      DO 80 K=1,3
         X(J,K)=XTEMP(1,J,K)
         V(J,K)=VTEMP(1,J,K)
   80 CONTINUE
      TIME=TIME+H
C
      CALL STORE
C
      IF(TIME.GE.TOTIME)GOTO 50
        IF(IG.GT.1000)THEN
          IG=1000
          GOTO 50
        ENDIF
      GOTO 25
C
   50 CALL OUT
```

```
C
      STOP
      END


      SUBROUTINE STORE
      DIMENSION CM(20),X(20,3),V(20,3),XSTORE(1000,20,3),
     +XTEMP(2,20,3),VTEMP(2,20,3),DELV(20,3),XT(20,3),VT(20,3)
      COMMON/A/X,V,TOL,H,TOTIME,DELV,XT,VT,NB,IST,TIME,IG,XTEMP,
     +VTEMP,CM
      COMMON/B/NORIG,XSTORE
      DO 21 J=1,NB
      DO 21 K=1,3
         X(J,K)=XTEMP(1,J,K)
         V(J,K)=VTEMP(1,J,K)
   21 CONTINUE
C
C UP TO 1000 POSITIONS ARE STORED. THESE ARE TAKEN EVERY 50 STEPS.
C
      IST=IST+1
      IF((IST/50)*50.NE.IST)GOTO 50
      IG=IST/50
      IF(IG.GT.1000)GOTO 50
        DO 22 J=1,NB
        DO 22 K=1,3
           XSTORE(IG,J,K)=X(J,K)
   22 CONTINUE
   50 RETURN
      END


      SUBROUTINE START
      DIMENSION CM(20),X(20,3),V(20,3),XSTORE(1000,20,3),
     +XTEMP(2,20,3),VTEMP(2,20,3),DELV(20,3),XT(20,3),VT(20,3)
      COMMON/A/X,V,TOL,H,TOTIME,DELV,XT,VT,NB,IST,TIME,IG,XTEMP,
     +VTEMP,CM
      COMMON/B/NORIG,XSTORE
      OPEN(UNIT=21,FILE='FAST.DAT')
      OPEN(UNIT=22,FILE='LAST.DAT')
C
C THE PROGRAMME AS PROVIDED IS FOR AN INVERSE-SQUARE LAW AND
C USES UNITS FOR WHICH THE UNIT MASS IS THAT OF THE SUN, THE
C THE UNIT OF DISTANCE IS THE ASTRONOMICAL UNIT (MEAN SUN-EARTH
C DISTANCE),THE UNIT OF TIME IS THE YEAR AND THE GRAVITATIONAL
C CONSTANT IS 4*PI**2.
C
      WRITE(6,'('' INPUT THE NUMBER OF BODIES'')')
          READ(5,*)NB
      WRITE(6,'('' INPUT THE VALUES OF CM IN SOLAR-MASS UNITS. '')')
      WRITE(6,'('' THE TROJAN ASTEROID MASSES CAN BE PUT AS ZERO'')')
```

```
      DO 1 I=1,NB
      WRITE(6,500)I
500 FORMAT(25H READ IN THE VALUE OF CM[,I3, 1H])
        READ(5,*)CM(I)
  1 CONTINUE
      WRITE(6,'('' INPUT THE INITIAL TIMESTEP [years]'')')
        READ(5,*)H
      WRITE(6,'('' INPUT TOTAL TIME FOR THE SIMULATION [years]'')')
        READ(5,*)TOTIME
C
C THE PROGRAMME ASKS THE USER TO SPECIFY A TOLERANCE, THE MAXIMUM
C ABSOLUTE ERROR THAT CAN BE TOLERATED IN ANY POSITIONAL COORDINATE
C (X, Y OR Z). IF THIS IS SET TOO LOW THEN THE PROGRAMME CAN BECOME
C VERY SLOW. FOR COMPUTATIONS INVOLVING PLANETS A TOLERANCE OF 1.0E-6
C (c. 150 KM) IS USUALLY SATISFACTORY.
C
      WRITE(6,'('' INPUT THE TOLERANCE '')')
      WRITE(6,'('' SEE COMMENT ABOVE THIS STATEMENT IN LISTING'')')
        READ(5,*)TOL
      WRITE(6,'('' THE CALCULATION CAN BE DONE RELATIVE TO AN '')')
      WRITE(6,'('' ARBITRARY ORIGIN OR WITH RESPECT TO ONE OF '')')
      WRITE(6,'('' THE BODIES AS ORIGIN. INPUT ZERO FOR AN '')')
      WRITE(6,'('' ARBITRARY ORIGIN OR THE NUMBER OF THE BODY.'')')
      WRITE(6,'('' IF A BODY IS CHOSEN AS ORIGIN THEN ALL ITS'')')
      WRITE(6,'('' POSITIONAL AND VELOCITY VALUES ARE SET TO ZERO'')')
        READ(5,*)NORIG
      DO 31 J=1,NB
      WRITE(6,100)J
100   FORMAT(23H INPUT [X,Y,Z] FOR BODY,I3)
        READ(5,*)X(J,1),X(J,2),X(J,3)
      WRITE(6,200)J
200 FORMAT(32H INPUT [XDOT,YDOT,ZDOT] FOR BODY,I3)
        READ(5,*)V(J,1),V(J,2),V(J,3)
 31 CONTINUE
      RETURN
      END


      SUBROUTINE ACC
      DIMENSION CM(20),X(20,3),V(20,3),XSTORE(1000,20,3),R(3),
     +XTEMP(2,20,3),VTEMP(2,20,3),DELV(20,3),XT(20,3),VT(20,3),DD(3)
      COMMON/A/X,V,TOL,H,TOTIME,DELV,XT,VT,NB,IST,TIME,IG,XTEMP,
     +VTEMP,CM
      COMMON/B/NORIG,XSTORE
C
C THE PROGRAMME AS PROVIDED IS FOR AN INVERSE-
C SQUARE LAW AND USES UNITS FOR WHICH THE UNIT MASS IS THAT OF THE SUN,
C THE UNIT OF DISTANCE IS THE ASTRONOMICAL UNIT (MEAN SUN-EARTH DISTANCE),
C THE UNIT OF TIME IS THE YEAR AND THE GRAVITATIONAL CONSTANT IS 4*PI**2.
C HOWEVER, THE USER MAY MODIFY THE SUBROUTINE "ACC" TO CHANGE TO ANY OTHER
```

```
C FORCE LAW AND/OR ANY OTHER SYSTEM OF UNITS.
C
C SET THE VALUE OF G IN ASTRONOMICAL UNITS
      PI=4.0*ATAN(1.0)
      G=4*PI*PI
      DO 1 J=1,NB
      DO 1 K=1,3
         DELV(J,K)=0
    1 CONTINUE
C THE FOLLOWING PAIR OF DO LOOPS FINDS INTERACTIONS FOR ALL PAIRS
C OF BODIES
      DO 2 J=1,NB-1
      DO 2 L=J+1,NB
      DO 3 K=1,3
         R(K)=XT(J,K)-XT(L,K)
    3 CONTINUE
      RRR=(R(1)**2+R(2)**2+R(3)**2)**1.5
      DO 4 K=1,3
C THE NEXT TWO STATEMENTS GIVE THE CONTRIBUTIONS TO THE THREE
C COMPONENTS OF ACCELERATION DUE TO BODY J ON BODY L AND DUE TO
C BODY L ON BODY J.
         DELV(J,K)=DELV(J,K)-G*CM(L)*R(K)/RRR
         DELV(L,K)=DELV(L,K)+G*CM(J)*R(K)/RRR
    4 CONTINUE
    2 CONTINUE
C IF ONE OF THE BODIES IS TO BE THE ORIGIN THEN ITS ACCELERATION IS
C SUBTRACTED FROM THAT OF ALL OTHER BODIES.
      IF(NORIG.EQ.0)GOTO 10
         DD(1)=DELV(NORIG,1)
         DD(2)=DELV(NORIG,2)
         DD(3)=DELV(NORIG,3)
      DO 6 J=1,NB
      DO 6 K=1,3
         DELV(J,K)=DELV(J,K)-DD(K)
    6 CONTINUE
   10 RETURN
      END


      SUBROUTINE OUT
      DIMENSION CM(20),X(20,3),V(20,3),XSTORE(1000,20,3),
     +XTEMP(2,20,3),VTEMP(2,20,3),DELV(20,3),XT(20,3),VT(20,3)
      COMMON/A/X,V,TOL,H,TOTIME,DELV,XT,VT,NB,IST,TIME,IG,XTEMP,
     +VTEMP,CM
      COMMON/B/NORIG,XSTORE
C  (X,Y) VALUES FOR THE LEADING ASTEROID ARE PLACED IN DATA FILE
C  FAST.DAT AND FOR THE FOLLOWING ASTEROID IN DATA FILE LAST.DAT.
C
C FOR THE TROJAN ASTEROID PROBLEM THE JUPITER RADIUS VECTOR IS
C ROTATED TO PUT IT ON THE Y AXIS. THE POSITIONS OF THE ASTEROIDS
```

```
C RELATIVE TO JUPITER ARE PLOTTED.
    LAST=MIN(IG,1000)
    DO 60 I=1,LAST
C THETA IS THE ANGLE BETWEEN THE X AXIS AND THE JUPITER RADIUS VECTOR
        THETA=ATAN2(XSTORE(I,2,2),XSTORE(I,2,1))
        XSTORE(I,2,2)=SQRT(XSTORE(I,2,2)**2+XSTORE(I,2,1)**2)
        XSTORE(I,2,1)=0
C NOW THE ASTEROID RADIUS VECTORS ARE ROTATED BY PI/2-THETA
    DO 61 J=1,2
        AA=XSTORE(I,J+2,1)*SIN(THETA)-XSTORE(I,J+2,2)*COS(THETA)
        BB=XSTORE(I,J+2,1)*COS(THETA)+XSTORE(I,J+2,2)*SIN(THETA)
        XSTORE(I,J+2,1)=AA
        XSTORE(I,J+2,2)=BB
 61 CONTINUE
 60 CONTINUE
C THE MODIFIED POSITIONS ARE NOW OUTPUT TO DATA FILES.
    DO 63 J=3,4
        N=18+J
        REWIND N
    DO 64 I=1,LAST
        WRITE(N,*)XSTORE(I,J,1),XSTORE(I,J,2)
 64 CONTINUE
 63 CONTINUE
    RETURN
    END
```

Physical Constants and Useful Data

Basic Units

Length	metre	m
Mass	kilogram	kg
Time	second	s
Current	ampere	A
Magnetic field	tesla	T
Absolute temperature	kelvin	K

Derived Units

Force	newton	$N = kg\,m\,s^{-2}$
Energy	joule	$J = N\,m$
Pressure	pascal	$P = N\,m^{-2}$
Power	watt	$W = J\,s^{-1}$
Electric charge	coulomb	$C = A\,s$
Electric potential	volt	$V = W\,A^{-1}$
Capacitance	farad	$F = C\,V^{-1}$
Inductance	henry	$H = V\,A^{-1}\,s$

Multiples and Submultiples of Units

peta	P	10^{15}	$Pm = 10^{15}\,m$
tera	T	10^{12}	$Tm = 10^{12}\,m$

(Continued)

(*Continued*)

giga	G	10^9	$Gm = 10^9$ m
mega	M	10^6	$Mm = 10^6$ m
kilo	k	10^3	$km = 10^3$ m (kilometre)
hecto	h	10^2	$hJ = 100$ J
deca	da	10	$daJ = 10$ J
deci	d	10^{-1}	$dJ = 10^{-1}$ J
centi	c	10^{-2}	$cm = 10^{-2}$ m (centimetre)
milli	m	10^{-3}	$mm = 10^{-3}$ m (millimetre)
micro	μ	10^{-6}	$\mu m = 10^{-6}$ m (micron)
nano	n	10^{-9}	$nm = 10^{-9}$ m (nanometre)
pico	p	10^{-12}	$pm = 10^{-12}$ m
femto	f	10^{-15}	$fm = 10^{-15}$ m

Physical Constants

Gravitational constant	$G = 6.6738 \times 10^{-11}$ m^2 kg^{-1} s^{-2}
Electron charge	$e = 1.6023 \times 10^{-19}$ C
Electron mass	$m_e = 9.1094 \times 10^{-31}$ kg
Proton mass	$m_p = 1.6726 \times 10^{-27}$ kg
Neutron mass	$m_n = 1.6749 \times 10^{-27}$ kg
Atomic mass unit	$amu = 1.66054 \times 10^{-27}$ kg
Boltzmann constant	$k = 1.3806 \times 10^{23}$ J K^{-1}
Planck constant	$h = 6.6261 \times 10^{-34}$ J s
$\hbar = h/2\pi$	$\hbar = 1.0546 \times 10^{-34}$ J s
Electron volt	$eV = 1.6022 \times 10^{-19}$ J
Speed of light	$c = 2.9979 \times 10^8$ m s^{-1}
Bohr magneton	$\mu_B = 9.2741 \times 10^{-24}$ J T^{-1}
Nuclear magneton	$\mu_N = 5.0508 \times 10^{-24}$ J T^{-1}
Astronomical unit	$au = 1.4960 \times 10^{11}$ m
Year	$a(yr) = 3.1557 \times 10^7$ s
Mass of Sun	$M_\odot = 1.9886 \times 10^{30}$ kg
Mass of Earth	$M_\oplus = 5.9722 \times 10^{24}$ kg
Mass of Moon	$M_m = 7.3477 \times 10^{22}$ kg

Solutions to Examples and Problems

Chapter 1

Exercises 1

1.1 The force per unit displacement $\kappa = 1/0.01 = 100.0$. Hence

$$\omega = \sqrt{\kappa/m} = \sqrt{100.0} = 10 \text{ radians s}^{-1}.$$

$$\text{Period } P = \frac{1}{n} = \frac{2\pi}{\omega} = 0.628 \text{ s}.$$

1.2 $25°$ in radians is $25\pi/180$.

From (1.9)$1 = 2\pi\sqrt{\dfrac{l}{9.8}}\left\{1 + \dfrac{1}{16}\left(\dfrac{25\pi}{180}\right)^2\right\} = 2.031\sqrt{l}.$

Hence $l = 1/2.031^2 = 0.242 \text{ m}.$

1.3 The kinetic energy at the equilibrium position equals the total energy. Hence

$$\frac{1}{2}mV^2 = \frac{1}{2}\kappa A^2 \quad \text{or} \quad V = \sqrt{\frac{\kappa}{m}}A = \sqrt{\frac{60}{0.5}} \times 0.05 = 0.548 \text{ m s}^{-1}.$$

1.4 From (1.18) the angular frequency is given by

$$\omega_l^2 = \omega^2 - \frac{f^2}{4m^2} = \frac{\kappa}{m} - \frac{f^2}{4m^2} = \frac{2}{0.1} - \frac{0.1^2}{4 \times 0.1^2} = 19.75.$$

The frequency is $n = \frac{\omega_l}{2\pi} = \frac{\sqrt{19.75}}{2\pi} = 0.707 \text{ Hz}.$

The reduction in amplitude is $\exp\left(-\frac{f}{2m}t\right) = \exp\left(-\frac{0.1}{0.2} \times 10\right) = 0.00674$.

1.5 For critical damping $f = 4m^2\omega^2 = 4m\kappa$ so that $\frac{f}{2m} = 2\kappa$. Hence in 5 seconds the amplitude will be $A(5) = A(0)\exp\left(-\frac{f}{2m}t\right) = 10\exp(-2 \times 0.5 \times 5) = 0.0674\,\text{cm}$.

1.6 From (1.29a) the steady state amplitude is given by

$$|x| = \frac{F}{\sqrt{m^2(\kappa/m - \omega_F^2)^2 + f^2\omega_F^2}}$$

$$= \frac{F}{\sqrt{0.01(10/0.1 - 25)^2 + 25}} = 0.111F.$$

The phase is given by $\phi = \tan^{-1}\left(\frac{-5}{0.1(10/0.1-25)}\right) = -33.7°$.

Note: The sine of $-33.6°$ is negative and the cosine positive.

Problems 1

1.1 If the cylinder has cross-section A then to support the weight of the piston

$$PA = mg.$$

The pressure of the gas is inversely proportional to the distance below the piston so that if the piston moves downwards through a small distance x the pressure changes to P' where

$$\frac{P'}{P} = \frac{h}{h-x} = \left(1 - \frac{x}{h}\right)^{-1} \approx 1 + \frac{x}{h}.$$

The net force on the piston (weight --- pressure) is now $F = mg - P'A$ or

$$F = mg - P\left(1 + \frac{x}{h}\right)A = -PA\frac{x}{h} = -\frac{mg}{h}x.$$

This gives the differential equation

$$\frac{d^2x}{dt^2} = -\frac{g}{h}x$$

corresponding to simple harmonic motion with frequency $\frac{1}{2\pi}\sqrt{\frac{g}{h}}$.

1.2

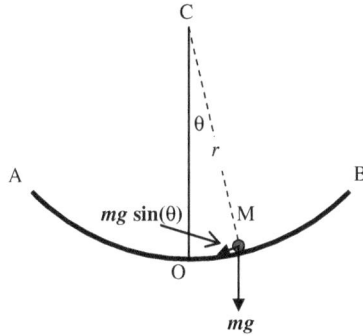

The component of the downward force due to gravity along the direction of motion is

$$F_g = -mg\sin(\theta) \approx -mg\theta \text{ if } \theta \text{ is small.}$$

The negative sign indicates that the force is in the opposite direction to the displacement.

The force due to friction is

$$F_V = -f\frac{dx}{dt} = -fr\frac{d\theta}{dt}$$

since the distance from O along the cylinder is $x = r\theta$.

The differential motion describing the motion is

$$m\frac{d^2x}{dt^2} = mr\frac{d^2\theta}{dt^2} = -fr\frac{d\theta}{dt} - mg\theta$$

or

$$\frac{d^2\theta}{dt^2} + \frac{f}{m}\frac{d\theta}{dt} + \frac{g}{r}\theta = 0.$$

Comparing this equation with (1.14) and (1.18) it is evident that the un-damped angular frequency, ω, is given by $\omega^2 = g/r$ and the lightly damped frequency by

$$\omega_f^2 = \frac{g}{r} - \frac{f^2}{4m^2}.$$

Hence the frequency is $n = \frac{\omega_f}{2\pi} = \frac{1}{2\pi}\sqrt{\frac{g}{r} - \frac{f^2}{4m^2}}$.

1.3

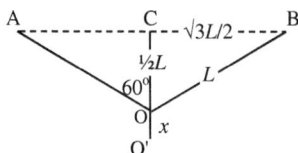

When the mass moves downwards by x each spring is stretched by an amount

$$\delta L = \text{O}'\text{B} - \text{OB} = \sqrt{\frac{3}{4}L^2 + \left(\frac{1}{2}L + x\right)^2} - L \approx \frac{1}{2}x$$

where, using the binomial theorem approximation $\text{O}'\text{B} \approx L + \frac{1}{2}x$. The vertical upward force exerted by each spring is

$$F_V = \frac{1}{2}\kappa x \cos(\text{C}\hat{\text{O}}'\text{B}) = \frac{1}{2}\kappa x \frac{\frac{1}{2}L + x}{L + \frac{1}{2}x}.$$

Using the standard binomial approximation we find

$$\frac{\frac{1}{2}L + x}{L + \frac{1}{2}x} = \frac{1}{2}\left(1 + \frac{3}{2}\frac{x}{L}\right).$$

When multiplied by $\frac{1}{2}\kappa x$, which we may think of as $\frac{1}{2}\kappa L \times x/L$ then, if we neglect terms in $(x/L)^2$,

$$F_V = \frac{1}{4}\kappa x.$$

Taking both springs into account the upward force is $\frac{1}{2}\kappa x$ and the angular frequency is

$$\omega = \sqrt{\frac{\kappa}{2m}} \quad \text{giving frequency } n = \frac{\omega}{2\pi} = \frac{1}{2\pi}\sqrt{\frac{\kappa}{2m}}.$$

Chapter 2

Exercises 2

2.1 The wavelength $\lambda = 2l = 39\,\text{m}$. Frequency $n = \frac{c}{\lambda} = \frac{300}{39} = 7.69\,\text{Hz}$.

2.2 From Equation (2.4) $n = \frac{1}{2 \times 0.2} \sqrt{\frac{10}{0.1}} = 25$ Hz. The second and third harmonics have frequencies 50 Hz and 75 Hz, respectively.

Chapter 3

Exercises 3

3.1 (a) $Z = (2 + 3) \, \Omega = 5\Omega$

(b) $\frac{1}{Z} = \left(\frac{1}{2} + \frac{1}{3}\right) S = \frac{5}{6}S$

Hence $Z = 1.2\Omega$.

3.2 $V \exp(i\pi) = V\{\cos(\pi) + i\sin(\pi)\} = -V$.

3.3 (a) $L = (2 + 5) \, H = 7 \, H$.

Impedance is $i|100 \times 7|\Omega = 700i \, \Omega$.

(b) $\frac{1}{L} = \left(\frac{1}{2} + \frac{1}{5}\right) H^{-1}$ or $L = \frac{10}{7} \, H$.

Impedance is $|i100 \times \frac{10}{7}|\Omega = \frac{1,000}{7}i \, \Omega$.

3.4 The impedances of the individual capacitors are

$z_1 = -\frac{i}{100 \times 10^{-3}} = -10i \, \Omega$ and $z_2 = -\frac{i}{100 \times 3 \times 10^{-3}} = -\frac{10}{3}i \, \Omega$.

(a) Impedance is $-i\left(10 + \frac{10}{3}\right) \Omega = -\frac{40}{3}i \, \Omega$.

(b) Impedance given by $\frac{1}{Z} = \frac{1}{z_1} + \frac{1}{z_2} = -\frac{1}{10i} - \frac{3}{10i} = -\frac{2}{5i} \, S$.

Hence

$$Z = -\frac{5}{2}i \, \Omega.$$

3.5 From (3.16) $Z = 5 + i\left(0.2 \times 100\pi - \frac{1}{100\pi \times 10^{-3}}\right) = 5 + 59.65i \, \Omega$.

Phase shift $= \tan^{-1}\left(\frac{59.65}{5}\right) = 85.2°$ (Note: Both $\sin(\delta)$ and $\cos(\delta)$ are positive.

3.6 Angular frequency $= (0.2 \times 2 \times 10^{-3})^{-1/2} = 50.0$ radians s^{-1}. The maximum impedance is given by (3.29) with $\omega = \omega_0$ and is

$$Z_{max} = \frac{R^2 + \omega_0^2 L^2}{R(1 + \omega_0^2 C^2 R^2)^{1/2}}$$

$$= \frac{1 + (50 \times 0.2)^2}{\{1 + (50 \times 2 \times 10^{-3})^2\}^{1/2}} = 100 \, \Omega.$$

Problems 3

3.1 For the top arm the impedance is

$$Z = 10 + i(0.4\omega - 10^6/\omega)$$

and the current is

$$I_1 = \frac{V}{Z} = V\frac{10 - i(0.4\omega - 10^6/\omega)}{100 + (0.4\omega - 10^6/\omega)^2}.$$

Similarly

$$I_2 = V\frac{2 - i(0.1\omega - 5 \times 10^5/\omega)}{4 + (0.1\omega - 5 \times 10^5/\omega)^2}.$$

The magnitude of the current in the main circuit is

$$I = |I_1 + I_2| = V(A^2 + B^2)^{1/2}$$

where

$$A = \frac{10}{100 + (0.4\omega - 10^6/\omega)^2} + \frac{2}{4 + (0.1\omega - 5 \times 10^5/\omega)^2}$$

and

$$B = \frac{0.4\omega - 10^6/\omega}{100 + (0.4\omega - 10^6/\omega)^2} + \frac{0.1\omega - 5 \times 10^5/\omega}{4 + (0.1\omega - 5 \times 10^5/\omega)^2}.$$

The following is a heavily commented FORTRAN77 program that calculates the main-circuit current for various frequencies. The pairs of values (frequency, current) are stored in the file LCRSERPAR.DAT.

```
      PROGRAM LCRSERPAR

C V IS THE AMPLITUDE OF THE APPLIED PD
C R(1) AND R(2) ARE THE RESISTANCES
C C(1) AND C(2) ARE THE CAPACITANCES
C L(1) AND L(2) ARE THE INDUCTANCES
C N IS THE FREQUENCY (Hz)
C W IS THE ANGULAR FREQUENCY
C RE(1) AND RE(2) ARE THE REAL PARTS OF THE
C CURRENTS
```

```
C IM(1) AND IM(2) ARE THE IMAGINARY PARTS OF THE
C CURRENTS
C REITOTAL IS THE REAL PART OF THE CURRENT IN
C THE MAIN CIRCUIT
C IMITOTAL IS THE IMAGINARY PART OF THE
C CURRENT IN THE MAIN CIRCUIT
C ITOTAL IS THE MAGNITUDE OF THE CURRENT IN
C THE MAIN CIRCUIT

      DIMENSION R(2),C(2),RE(2)
      REAL L(2),IM(2),N,IMITOTAL,ITOTAL

C SET VALUES OF VARIABLES

      DATA V/10.0/
      DATA R/10.0,2.0/
      DATA C/1.0E-6,2.0E-6/
      DATA L/4.0E-1,1.0E-1/

C SET VALUE OF TWO X PI

      TPI=8.0*ATAN(1.0)

C OPEN FILE FOR STORING VALUES OF N AND ITOTAL

      OPEN(UNIT=10,FILE='LCRSERPAR.DAT')

C FREQUENCIES FROM 10 TO 500

      DO 1 N=10,500
        W= N*TPI
      DO 2 J=1,2
        RE(J)=R(J)/(R(J)**2+(W*L(J)-1.0/W/C(J))**2)*V
        IM(J)=(1.0/W/C(J)-W*L(J))/(R(J)**2+(W*L(J)-
     1.0/W/C(J))**2)*V
2 CONTINUE
      REITOTAL=RE(1)+RE(2)
      IMITOTAL=IM(1)+IM(2)
      ITOTAL=SQRT(REITOTAL**2+IMITOTAL**2)
      WRITE(10,*)N,ITOTAL
```

1 CONTINUE
STOP
END

The plot from LCRSERPAR.DAT shows two peaks of current corresponding to resonances in the two arms of the circuit.

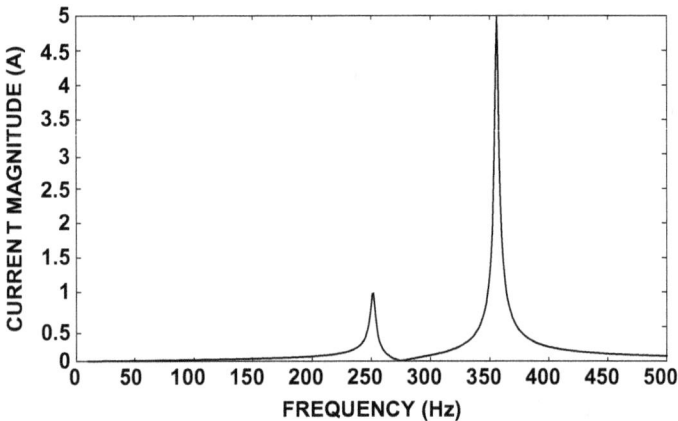

3.2 From (3.26)

$$Z = \frac{R^2 + \omega^2 L^2}{R + i\omega\{C(R^2 + \omega^2 L^2) - L\}}$$

$$= \frac{(R^2 + \omega^2 L^2)|R - i\omega\{C(R^2 + \omega^2 L^2) - L\}|}{R^2 + \omega^2\{C(R^2 + \omega^2 L^2) - L\}^2}.$$

The real and imaginary parts of the impedance are

$$\text{Re}(Z) = \frac{R(R^2 + \omega^2 L^2)}{R^2 + \omega^2\{C(R^2 + \omega^2 L^2) - L\}^2}$$

and

$$\text{Im}(Z) = \frac{\omega^2\{L - C(R^2 + \omega^2 L^2)\}}{R^2 + \omega^2\{C(R^2 + \omega^2 L^2) - L\}^2}.$$

The magnitude of the total impedance is given by

$$Z_T = \{(Re(Z_1) + Re(Z_2))^2 + (Im(Z_1) + Im(Z_2))^2\}^{1/2}.$$

The following is a heavily commented FORTRAN77 program that calculates the total impedance for various frequencies. The pairs oF values (frequency, impedance) are stored in the file LCRPARSER.DAT.

The plot from LCRPARSER.DAT shows two peaks of impedance corresponding to resonances in the two components of the circuit.

```
PROGRAM LCRPARSER

C R(1) AND R(2) ARE THE RESISTANCES
C C(1) AND C(2) ARE THE CAPACITANCES
C L(1) AND L(2) ARE THE INDUCTANCES
C N IS THE FREQUENCY (Hz)
C W IS THE ANGULAR FREQUENCY
C RE(1) AND RE(2) ARE THE REAL PARTS OF THE
C IMPEDANCES
C IM(1) AND IM(2) ARE THE IMAGINARY PARTS OF THE
C IMPEDANCES
C REITOTAL IS THE REAL PART OF THE TOTAL
C IMPEDANCE
C IMITOTAL IS THE IMAGINARY PART OF THE TOTAL
C IMPEDANCE
C ITOTAL IS THE MAGNITUDE OF THE TOTAL
C IMPEDANCE

      DIMENSION R(2),C(2),RE(2)
      REAL L(2),IM(2),N,IMITOTAL,ITOTAL

C SET VALUES OF VARIABLES

      DATA R/10.0,2.0/
      DATA C/1.0E-6,2.0E-6/
      DATA L/4.0E-1,1.0E-1/
```

```
C SET VALUE OF TWO X PI

  TPI=8.0*ATAN(1.0)

C OPEN FILE FOR STORING VALUES OF N AND ITOTAL

  OPEN(UNIT=10,FILE='LCRPARSER.DAT')

C FREQUENCIES FROM 10 TO 500

  DO 1 N=10,500
    W=N*TPI
  DO 2 J=1,2
    DIVISOR=R(J)**2+W*W*(C(J)*(R(J)**2+W*W*L
      (J)**2)-L(J))**2
    RE(J)=R(J)*(R(J)**2+W*W*L(J)**2)/DIVISOR
    IM(J)=W*(L(J)-C(J)*(R(J)**2+W*W*L(J)**2))/
      DIVISOR
  2 CONTINUE
    REITOTAL=RE(1)+RE(2)
    IMITOTAL=IM(1)+IM(2)
    ITOTAL=SQRT(REITOTAL**2+IMITOTAL**2)
    WRITE(10,*)N,ITOTAL
  1 CONTINUE
    STOP
    END
```

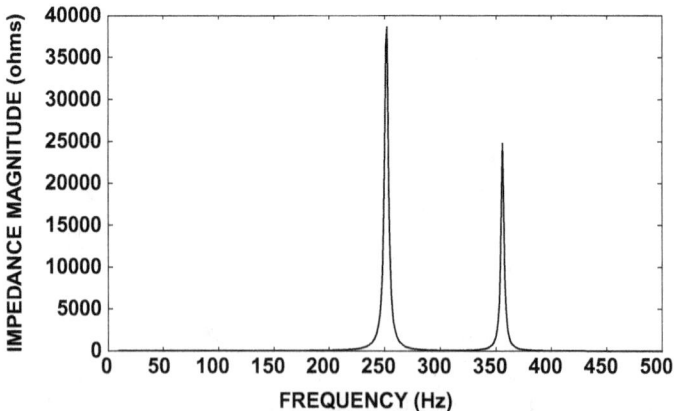

Chapter 4

Exercises 4

4.1 The relationship is equivalent to $\log r_n = n \log b + \log a$.

Planet	n	r_n	$\log r_n$	Formula
Mercury	0	0.387	−0.412	0.361
Venus	1	0.723	−0.141	0.626
Earth	2	1.000	0.000	1.085
Mars	3	1.524	0.183	1.882
Jupiter	5	5.203	0.716	5.659
Saturn	6	9.539	0.980	9.813
Uranus	7	19.19	1.283	17.02
Neptune	8	30.07	1.478	29.51

Visual estimation of the best straight line will vary from one individual to another. For comparison the analytical least-squares line, shown on the graph, gives $\log b = 0.239$ and $\log a = -0.443$. This gives the best relationship of this type as

$$r_n = 10^a \times (10^b)^n$$

or

$$r_n = 0.361 \times 1.734^n,$$

which gives the values in the final column of the table.

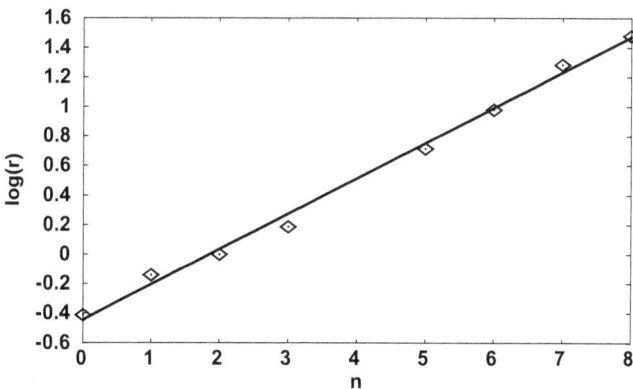

4.2 For solar-system units the mass of the Sun is 1 and $G = 4\pi^2$. This gives $P = r^{3/2}$ where r is in au and P in years. Hence for $r = 2.71$ au $P = 4.461$ years and for $r = 3.04$ au $P = 5.300$ years.

$$\text{For 2.71 au} \quad \frac{P_r}{P_J} = \frac{4.461}{11.862} = 0.376 \approx \frac{3}{8}(0.375).$$

$$\text{For 3.04 au} \quad \frac{P_r}{P_J} = \frac{5.300}{11.862} = 0.447 \approx \frac{4}{9}(0.444).$$

4.3 With all other factors unchanged the heating rate $W \propto e/R^{15/2}$. Hence the change in the heating rate if by a factor

$$f = \frac{0.02}{0.004}\frac{1}{1.1^{15/2}} = 2.45.$$

4.4 The minimum distance of Saturn from Uranus is 9.65 au and that of Jupiter from Uranus is 13.99 au. Hence the ratio of the maximum gravitational force on Uranus due to Jupiter to that die to Saturn is

$$\frac{F_J}{F_S} = \frac{317.8}{95.2}\left(\frac{9.65}{13.99}\right)^2 = 1.59.$$

Taken in conjunction with Jupiter's shorter orbital period this shows that Jupiter has more closest approaches to Uranus per unit time than does Saturn and, additionally, each closest approach has a larger perturbing effect.

From the relationship between period and orbital radius given in Exercise 4.2,

$$\frac{P_U}{P_J} = \left(\frac{19.19}{5.20}\right)^{3/2} = 7.09.$$

This departs from a 7:1 commensurability by 1.2%. The departure for Saturn:Jupiter is 0.7% and that for Neptune:Uranus is 1.9%.

4.5 The ratios of the periods of Table 4.1 satellites to those of Rhea are:

Mimas	$0.9422/4.5182 = 0.209$	Very dubious 1:5 (0.200) but not much further from 2:9 (0.222).
Enceladus	$1.3702/4.5182 = 0.303$	A 1% departure from 3:10.
Tethys	$1.8878/4.5182 = 0.418$	No obvious commensurability.
Dione	$2.7369/4.5182 = 0.606$	A 1% departure from 3:5.
Titan	$15.5492/4.5182 = 3.441$	No obvious commensurability.
Hyperion	$21.2766/4.5182 = 4.709$	No obvious commensurability.

Problems 4

4.1 Equation (4.11b) in its general form is

$$W = \frac{27}{16} \frac{(GM_P)^{5/2} M_S^2 re}{R^{15/2} QY}$$

where subscript S refers to the satellite and P to the planet. For the Earth–Moon system:

$M_{\text{Moon}} = 7.348 \times 10^{22}\,\text{kg}$, $M_{\text{Earth}} = 5.972 \times 10^{24}\,\text{kg}$, $r_{\text{Moon}} = 1.738 \times 10^6\,\text{m}$ $e_{\text{Moon}} = 0.056$, $R_{\text{Moon}} = 3.848 \times 10^8\,\text{m}$.

We also take $Q = 500$ and $Y = 10^{11}\,\text{Nm}^{-2}$. This gives

$$W = \frac{27}{16} \frac{(6.674 \times 10^{-11} \times 5.972 \times 10^{24})^{5/2} (7.348 \times 10^{22})^2 \times 1.738 \times 10^6 \times 0.056}{(3.848 \times 10^8)^{15/2} \times 500 \times 10^{11}}$$

$$= 2.3 \times 10^9\,\text{W}.$$

4.2 (i)

$$V_{Teth} = \left(\frac{GM_{sat}}{R_{Teth}}\right)^{1/2}$$

$$= \left(\frac{6.674 \times 10^{-11} \times 5.685 \times 10^{26}}{2.95 \times 10^8}\right)^{1/2}\,\text{m s}^{-1}$$

$$\doteq 1.134 \times 10^4\,\text{m s}^{-1} = 2.392\,\text{au year}^{-1}.$$

(ii) For the trailing satellite

$x = 2.95 \times 10^8 \times \cos 30° \, \text{m} = 2.555 \times 10^8 \, \text{m} = 1.708 \times 10^{-3}$ au.

$y = 2.95 \times 10^8 \times \sin 30° \, \text{m} = 1.475 \times 10^8 \, \text{m} = 9.860 \times 10^{-4}$ au.

For the leading satellite

$x = -1.708 \times 10^{-3}$ au, $y = 9.860 \times 10^{-4}$ au.

(iii) For the trailing satellite

$V_x = -2.392 \times \cos 60° = -1.196$ au year^{-1}

$V_y = 2.392 \times \sin 60°$ au year$^{-1} = 2.072$ au year^{-1}.

For the leading satellite

$V_x = -1.196$ au year^{-1}, $V_y = 2.072$ year^{-1}.

(iv) The table corresponding to Table 4.2 is:

	Saturn	Tethys	Satellite 1	Satellite 2
Mass	2.858×10^{-4}	3.80×10^{-10}	0	0
x	0	0	-1.708×10^{-3}	1.708×10^{-3}
y	0	1.972×10^{-3}	9.860×10^{-4}	9.860×10^{-4}
z	0	0	0	0
V_x	0	-2.392	-1.196	-1.196
V_y	0	0	-2.072	2.072
V_z	0	0	0	0

Running TROJANS and displaying the output gives

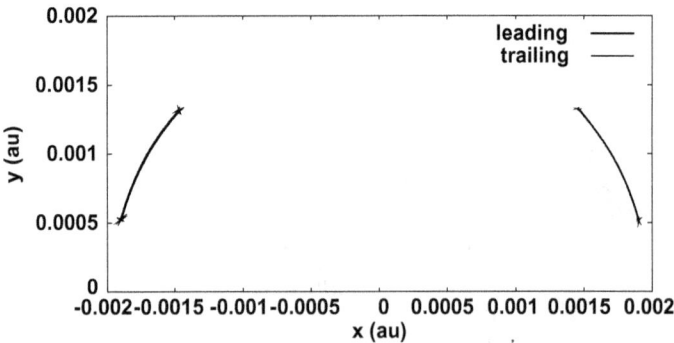

The figure show the large-scale view of the satellite paths. Because of the small mass of Tethys relative to Saturn the satellites are very loosely bound and oscillate about the mean position with large amplitude. A larger-scale view of part of the paths shows that they are in a state of oscillation about a mean position.

Chapter 5

Exercises 5

5.1 For a general isotope $^A_Z Q$ the number of protons is Z and the number of neutrons is A−Z. Hence for the given isotopes:

Isotope	Protons	Neutrons
$^{235}_{92}U$	92	143
$^{238}_{92}U$	92	146
$^{204}_{82}Pb$	82	122
$^{154}_{62}Sm$	62	92

5.2 From (5.2) the magnetic moment is

$$\mu = \gamma \times \frac{1}{2}\hbar = 2.517 \times 10^8 \times 0.5 \times 1.055 \times 10^{-34}$$

$$= 1.328 \times 10^{-26} \, \text{kg m}^2 \, \text{s}^{-2} \, \text{T}^{-1}.$$

This can also be expressed as 1.328×10^{-26} J T^{-1}, where J, the joule, is the SI unit of energy.

5.3 From Exercise 5.2, $\mu = 1.328 \times 10^{-26}$ J T^{-1}. Hence the energy of the spin-up state is

$$E_{up} = -1.328 \times 10^{-26} \times 3 = 3.984 \times 10^{-26} \, \text{J}$$

and E_{down} is the negative of that value. Hence

$$\frac{P(E_{up})}{P(E_{down})} = \frac{\exp\left(\frac{3.984 \times 10^{-26}}{1.381 \times 10^{-23} \times 100}\right)}{\exp\left(-\frac{3.984 \times 10^{-26}}{1.381 \times 10^{-23} \times 100}\right)} = \frac{1.000028849}{0.999971151}.$$

This gives

$$P(E_{up}) = \frac{1.000028849}{1.000028849 + 0.999971151} = 0.50001442$$

and

$$P(E_{down}) = 1 - P(E_{up}) = 0.49998558.$$

5.4 The Larmor frequency is from (5.16)

$$\nu = \frac{\omega}{2\pi} = \frac{\gamma B}{2\pi} = \frac{2.517 \times 10^8 \times 2}{2\pi} = 8.012 \times 10^7 \, \text{Hz}.$$

5.5 The ratio of the two signals is

$$\frac{S(A)}{S(B)} = \frac{1.2 \exp(-t/500)}{\exp(-t/900)}$$

where t is in milliseconds.

For $t = 100 S(A)/S(B) = 1.098$ and for $t = 900 S(A)/S(B) = 0.539$.

5.6 For a general value $u = a$

$$F(a) = \int_{-1}^{1} \exp(-2\pi i a x) dx$$

$$= \int_{-1}^{1} \cos(2\pi a x) dx - i \int_{-1}^{1} \sin(2\pi a x) dx$$

$$= \frac{1}{2\pi a} |\sin(2\pi a x)|_{-1}^{1} + \frac{1}{2\pi a} |\cos(2\pi a x)|_{-1}^{1}.$$

Because cosine is an even function the second term is zero and

$$F(a) = \frac{\sin(2\pi a)}{\pi a}.$$

For $a = 0.1$

$$F(0.1) = \frac{\sin(0.2\pi)}{0.1\pi}$$

and for $a = -0.1$

$$F(-0.1) = \frac{\sin(-0.2\pi)}{-0.1\pi} = \frac{\sin(0.2\pi)}{0.1\pi}.$$

Hence the FT is the same for both values of a. To determine the phase we note that B $(\sin(\phi))$ in (5.21) is zero so that ϕ is either 0 or π. However, since A $(\cos(\phi))$ is positive this indicates that $\phi = 0$.

5.7 Along the working region the z- field varies from 2 T to 2.004 T. From (5.16) the variation of Larmor frequency is from

$$\nu_1 = \frac{2.6751 \times 10^8 \times 2}{2\pi} = 8.5151 \times 10^7 \text{ Hz to}$$

$$\nu_2 = \frac{2.6751 \times 10^8 \times 2.004}{2\pi} = 8.5321 \times 10^7 \text{ Hz .}$$

Problems 5

5.1 From (5.6) the components of dipole moment in the field direction are

$$\frac{3}{2}g\mu_N, \quad \frac{1}{2}g\mu_N, \quad -\frac{1}{2}g\mu_N \quad \text{and} \quad -\frac{3}{2}g\mu_N$$

and from (5.7) the energies are

$$-\frac{3}{2}g\mu_N B, \quad -\frac{1}{2}g\mu_N B, \quad \frac{1}{2}g\mu_N B \quad \text{and} \quad \frac{3}{2}g\mu_N B.$$

Evaluating these with $g = 5.722$, $\mu_N - 5.0508 \times 10^{-27}$ J T^{-1} and $B = 10$ T gives

$$E_{3/2} = -4.3351 \times 10^{-25} \text{ J}, \quad E_{1/2} = -1.4450 \times 10^{-25} \text{ J}$$

$$E_{-3/2} = 4.3351 \times 10^{-25} \text{ J}, \quad E_{-1/2} = 1.4450 \times 10^{-25} \text{ J}.$$

The values of $\exp\{-E_m/(kT)\}$ are; respectively,

$$1.00314, \quad 1.00105, \quad 0.99895 \text{ and } 0.99687$$

giving $\sum_{-3/2}^{3/2} \exp\{E_m/(kT)\} = 4.00001$.

From (5.9) this gives probabilities

$$P(3/2) = 0.25078, \quad P(1/2) = 0.25026.$$

$$P(-1/2) = 0.24973 \quad \text{and} \quad P(-3/2) = 0.24922.$$

5.2 If the total collecting time is τ then the cumulative signal is proportional to

$$S = \rho \int_0^\tau \exp\left(-\frac{t}{T}\right) dt = \rho T \left\{1 - \exp\left(-\frac{\tau}{T}\right)\right\}$$

where ρ is relative proton density.

(i) $t = 400$ ms.

For T1

$$S(\text{A}) = 500 \times \left\{1 - \exp\left(-\frac{400}{500}\right)\right\} = 275.3$$

$$S(\text{B}) = 0.95 \times 900 \times \left\{1 - \exp\left(-\frac{400}{900}\right)\right\} = 306.8,$$

giving $S(\text{A})/S/(\text{B}) = 0.897$.

For T2

$$S(\text{A}) = 80 \times \left\{1 - \exp\left(-\frac{400}{80}\right)\right\} = 79.46$$

$$S(B) = 0.95 \times 60 \times \left\{1 - \exp\left(-\frac{400}{60}\right)\right\} = 56.93,$$

giving $S(\text{A})/S/(\text{B}) = 1.396$.

For T–3T2

$$\frac{S(\text{A})}{S(\text{B})} = \frac{275.3 - 3 \times 79.46}{306.8 - 3 \times 56.93} = 0.257.$$

This enhances tissue B relative to A.

(ii) $t = 100$ ms

For T1

$$S(\text{A}) = 500 \times \left\{1 - \exp\left(-\frac{100}{500}\right)\right\} = 90.63$$

$$S(\text{B}) = 0.95 \times 900 \times \left\{1 - \exp\left(-\frac{100}{900}\right)\right\} = 89.91,$$

giving $S(\text{A})/S/(\text{B}) = 1.008$.

For T2

$$S(A) = 80 \times \left\{1 - \exp\left(-\frac{100}{80}\right)\right\} = 57.08$$

$$S(B) = 0.95 \times 60 \times \left\{1 - \exp\left(-\frac{100}{60}\right)\right\} = 46.23,$$

giving $S(A)/S(B) = 1.235$.

For T2–0.95T2(B)

$$\frac{S(A)}{S(B)} = \frac{57.08 - 0.95 \times 46.23}{46.23 - 0.95 \times 46.23} = 5.69.$$

This enhances tissue A relative to B.

Chapter 6

Exercises 6

6.1 (a) $1s^2 2s^2 2p^1$ (b) $1s^2 2s^2 2p^5$.

6.2

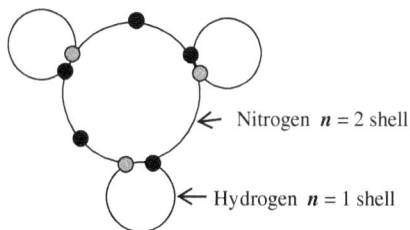

← Nitrogen $n = 2$ shell

← Hydrogen $n = 1$ shell

6.3

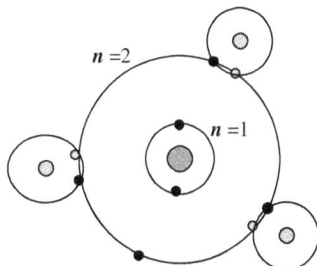

$n = 2$

$n = 1$

6.4 The difference of energy is

$$\Delta E = g_e \mu_B B = 2.0023 \times 9.274 \times 10^{-24} \times 0.5 = 9.285 \times 10^{-24} \text{ J}.$$

6.5

$$\frac{P_{1/2}}{P_{-1/2}} = \exp\left(\frac{\Delta E}{kT}\right) = \exp\left(\frac{9.285 \times 10^{-24}}{1.381 \times 10^{-23} \times 5}\right) = 1.144.$$

6.6

$$f(B) = \frac{W}{(B - B_0)^2 + W^2}$$

$$\frac{df}{dB} = -\frac{2W(B - B_0)}{\{(B - B_0)^2 + W^2\}^2}.$$

6.7 From Equations (6.12a) and (6.12b) the two peaks are at

$$B = \frac{h\nu}{g_e \mu_B} \pm \frac{a}{2g_e \mu_B}$$

$$= \frac{6.6261 \times 10^{-34} \times 9 \times 10^9}{2.0023 \times 9.2740 \times 10^{-24}} \pm \frac{5 \times 10^{-27}}{2 \times 2.0023 \times 9.2740 \times 10^{-24}}$$

$$= 0.32115 \pm 0.00013 \text{ T}.$$

Hence the two peaks are at 0.32128 T and 0.32102 T.

6.8 The possible combined spins of the two nuclei are given in the table below:

m_1	m_2	Σm
1	1	2
1	0	1
1	−1	0
0	1	1
0	0	0
0	−1	−1
−1	1	0
−1	0	−1
−1	−1	−2

There are five peaks in the spectrum. Corresponding to $\Sigma m = -2, -1, 0, 1$ and 2 with intensities in the ratio 1:2:3:2:1.

Problems 6

6.1 The following table gives the combinations of m_1, m_2 and m_3 for the three nuclei:

			m_3	
m_1	m_2	1	0	−1
1	1	**3**	**2**	**1**
1	0	**2**	**1**	**0**
1	−1	**1**	**0**	**−1**
0	1	**2**	**1**	**0**
0	0	**1**	**0**	**−1**
0	−1	**0**	**−1**	**−2**
−1	1	**1**	**0**	**−1**
−1	0	**0**	**−1**	**−2**
−1	−1	**−1**	**−2**	**−3**

The bold entries in the table are the 27 possible values of Σm. There are clearly seven peaks for which $-3 \leq \Sigma m \leq +3$ and the peak intensities are in the ratio 1:3:6:7:6:3:1.

6.2 From the result of Problem 6.1 for the black nuclei alone there would be seven peaks with intensities in the ratio 1:3:6:7:6:3:1. These peaks, represented as lines, would appear as follows:

From Table 6.3, for the white nuclei alone, for three equivalent spin-$\frac{1}{2}$ nuclei there are four peaks with intensities in the ratio 1:3:3:1. The total pattern consists of four copies of the black-nuclei peaks, in the appropriate ratios with separation 5.5 times the separations of the black-nuclei peaks. This is shown below.

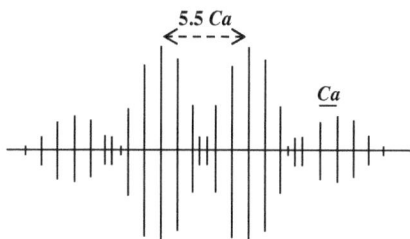

Chapter 7

Exercises 7

7.1 Energy of photon is $(E_2 - E_1) = (-3.401 + 13.605) \times 1.6022 \times 10^{-19}$ J $= 1.6349 \times 10^{-18}$ J

$$\lambda = \frac{c}{\nu} = \frac{ch}{E_2 - E_1}$$

$$= \frac{2.998 \times 10^8 \times 6.626 \times 10^{-34}}{1.6349 \times 10^{-18}} = 1.215 \times 10^{-7}\,\text{m}.$$

This is $121.5\,\text{nm}$ and is ultraviolet radiation.

7.2 From (7.11)

$$\frac{A_{21}}{B_{21}} = \frac{2h\nu^3}{c^2} = \frac{2 \times 6.626 \times 10^{-34} \times (6 \times 10^{14})^3}{(2.998 \times 10^8)^2} = 3.18 \times 10^{-6}.$$

From (7.8)

$$\frac{N_1}{N_2} = \frac{g_1}{g_2} \exp\left(\frac{h\nu}{kT}\right)$$

$$= \exp\left(\frac{6.626 \circ 10^{-34} \times 6 \times 10^{14}}{1.381 \times 10^{-23} \times 1000}\right) = 3.18 \times 10^{12}.$$

7.3 The distance the radar waves travel in $53\,\mu s$ is $2.998 \times 10^8 \times 53 \times 10^{-6}\,\text{m} = 15.89\,\text{km}$, giving an aircraft distance $7.94\,\text{km}$. Since the distance is known to be less than $30\,\text{km}$ there is no ambiguity in the distance. For a PRF of $2,000\,\text{Hz}$ the time between pulses is $0.0005\,\text{s}$ and if the return pulse was $(53 \times 10^{-6} + 0.0005)\,\text{s}$ behind the outward pulse the distance would be much greater than $30\,\text{km}$.

7.4 The total energy emitted in each pulse is $E_p = 0.5 \times 10^{-6} \times 5 \times 10^6 = 2.5\,\text{J}$. The average power transmitted is the energy per second or $P = 2.5 \times 1000 = 2,500\,\text{W}$.

Problems 7

7.1 From Equations (7.1) and (7.2)

$$E_n = \frac{hc}{\lambda} + E_1.$$

Calculating energy in units of eV

$$E_2 = \frac{6.626 \times 10^{-34} \times 2.998 \times 10^8}{1.215 \times 10^{-7} \times 1.602 \times 10^{-19}} - 13.6 = -3.4\,\mathrm{eV} = -\frac{E_1}{2^2}$$

$$E_3 = \frac{6.626 \times 10^{-34} \times 2.998 \times 10^8}{1.025 \times 10^{-7} \times 1.602 \times 10^{-19}} - 13.6 = -1.5\,\mathrm{eV} = -\frac{E_1}{3^2}$$

$$E_4 = \frac{6.626 \times 10^{-34} \times 2.998 \times 10^8}{9.72 \times 10^{-8} \times 1.602 \times 10^{-19}} - 13.6 = 0.84\,\mathrm{eV} \approx -\frac{E_1}{4^2}.$$

7.2 For the first set of pulses the time between pulses is 10^{-3} s so the possible there-and-back travel times for the pulse are $(2 \times 10^{-4} + n \times 10^{-3})$ s. This gives possible target distances

$$d_n = \frac{1}{2}c(2 \times 10^{-4} + n \times 10^{-3})\,\mathrm{m} \quad \text{with } n = 0, 1, 2 \text{ etc.}$$

In kilometres these are

$$d_0 = 29.98\,\mathrm{km}, \quad d_1 = 179.88\,\mathrm{km}, \quad d_2 = 329.78\,\mathrm{km},$$
$$d_3 = 479.68\,\mathrm{km} \quad \text{etc.}$$

For the second set of pulses the time between pulses is 7.692×10^{-4} s and the possible target distances are

$$d_n = \frac{1}{2}c(1.23 \times 10^{-4} + n \times 7.692 \times 10^{-4})\,\mathrm{m}.$$

In kilometres these are

$$d_0 = 18.44\,\mathrm{km}, \quad d_1 = 133.74\,\mathrm{km}, \quad d_2 = 249.04\,\mathrm{km},$$
$$d_3 = 364.35\,\mathrm{km}, \quad d_4 = 479.65\,\mathrm{km}.$$

The common distance is 479.7 km, which is the target distance.

7.3 From (7.19) $g = 7.5 \times 10^{-15} \times 248.5 = 1.864 \times 10^{-12}\,\mathrm{kg\,s^{-1}}$.
The frequency of the K absorption edge for copper is

$$\nu_0 = \frac{c}{\lambda_0} = \frac{2.998 \times 10^8}{1.3808 \times 10^{-10}} = 2.1712 \times 10^{18}\,\mathrm{Hz}$$

and

$$\omega_0 = 2\pi\nu_0 = 1.3642 \times 10^{19}\mathrm{s^{-1}}.$$

We also have k, as used in (7.17a) and (7.17b), given by

$$k = \frac{\omega_0}{\omega} = \frac{\lambda}{\lambda_0} = \frac{0.13750}{0.13808} = 0.99580.$$

From (7.17a)

$$\varepsilon' = -\frac{(9.109 \times 10^{-31})^2 \times (0.99580^2 - 1)}{(9.109 \times 10^{-31})^2(0.99580^2 - 1)^2 + (1.864 \times 10^{-12} \times 0.99580/1.3642 \times 10^{19})^2} - 1$$

$$= -0.375$$

and, from (7.17b),

$$\varepsilon'' = \frac{9.109 \times 10^{-31} \times 1.864 \times 10^{-12} \times 0.99580/1.3642 \times 10^{19}}{(9.109 \times 10^{-31})^2(0.99580^2 - 1)^2 + (1.864 \times 10^{-12} \times 0.99580/1.3642 \times 10^{19})^2}$$

$$= 6.674.$$

Chapter 8

Exercises 8

8.1 Assuming that the speed is well below the speed of light

$$\frac{1}{2}mv^2 = E$$

or

$$\frac{v}{c} = \frac{1}{c}\left(\frac{2E}{m}\right)^{1/2}$$

$$= \frac{1}{2.998 \times 10^8}\left(\frac{2 \times 7.1 \times 10^5 \times 1.602 \times 10^{-19}}{1.673 \times 10^{-27}}\right)^{1/2}$$

$$= 0.0389.$$

8.2 (i) The cyclotron frequency is given by

$$f = \frac{qB}{2\pi m} = \frac{1.602 \times 10^{-19} \times 1}{2\pi \times 1.673 \times 10^{-27}} = 1.524 \times 10^7 \text{ Hz.}$$

(ii) From (8.6)

$$r = \frac{\sqrt{2mE}}{qB}$$

$$= \frac{\sqrt{2 \times 1.673 \times 10^{-27} \times 2 \times 10^6 \times 1.602 \times 10^{-19}}}{1.602 \times 10^{-19} \times 1}$$

$$= 0.204 \text{ m.}$$

8.3 If $f/f_0 \geq 0.9999$ then $m_0/m \geq 0.9999$. From (8.7)

$$\sqrt{1 - v_{max}^2/c^2} \geq 0.9999$$

or

$$v_{max} = \sqrt{1 - 0.9999^2}c = 0.01414c.$$

8.4 (i) For an electron $m_0c^2 = 9.109 \times 10^{-31} \times (2.998 \times 10^8)^2 \div 1.602 \times 10^{-19} = 511.1 \text{ eV}$. From (8.13), on entry the electron speed is

$$v = c\left\{1 - \left(\frac{m_0c^2}{E + m_0c^2}\right)^2\right\}^{1/2}$$

$$= 2.998 \times 10^8 \left\{1 - \left(\frac{511.1}{10^3 + 511.1}\right)^2\right\}^{1/2}$$

$$= 2.821 \times 10^8 \text{ m s.}$$

On exit the speed is

$$v = c\left\{1 - \left(\frac{m_0c^2}{E + m_0c^2}\right)^2\right\}^{1/2}$$

$$= 2.998 \times 10^8 \left\{1 - \left(\frac{511.1}{10^{10} + 511.1}\right)^2\right\}^{1/2}$$

$$= 2.998 \times 10^8 \text{ m s}^{-1}.$$

(ii) The length of the electrode is the distance travelled by the electrons in 10^{-9} s. On entry this is

$$d_{entry} = 2.821 \times 10^8 \times 10^{-9} = 0.2821 \, \text{m} = 28.21 \, \text{cm}.$$

Similarly, on exit, length of the electrode is 29.98 cm.

8.5 Since there are 48 bending magnets, each must bend the electron path by $\pi/24$ radians (7.5°) so this is the horizontal spread. The vertical spread is given by (8.14) and is

$$\phi = \frac{m_e c^2}{E} = \frac{511.1}{3 \times 10^9} = 1.70 \times 10^{-7} \, \text{radians} = 0.035 \, \text{seconds of arc}.$$

8.6 (i) The particle will be travelling at almost the speed of light so the time for one circuit is

$$t = \frac{2.7 \times 10^4}{2.998 \times 10^8} = 9.00 \times 10^{-5} \, \text{s}.$$

(ii) From (8.15)

$$\lambda_0 = \frac{1.864 \times 10^{-9}}{BE^2} = \frac{1.864 \times 10^{-9}}{8.4 \times (7 \times 10^3)^2} = 4.53 \times 10^{-18} \, \text{m}.$$

This is in the γ-ray region of the electromagnetic spectrum.

Problems 8

8.1 From (8.4) the speed of the emitted particles is

$$v_R = \frac{qBR}{m}$$

where R is the radius of the accelerating chamber. The minimum particle mass, m_{min}, to ensure that the emitted particles have speed no greater than v_{max} is given by

$$m_{min} = \frac{qBR}{v_{max}} = \frac{1.602 \times 10^{-19} \times 1.5 \times 5}{0.05 \times 2.998 \times 10^8} = 8.02 \times 10^{-26}.$$

This is $8.02 \times 10^{-26}/1.661 \times 10^{-27} = 48.3$ atomic mass units.

The energy of the emitted particles is, from (8.6)

$$E = \frac{(qBR)^2}{2m},$$

so the mass should be as small as possible consistent with being greater than $m_{\min}$. Hence iron ions should be used. The energy, in eV, is then

$$E = \frac{(1.602 \times 10^{-19} \times 1.5 \times 5)^2}{56 \times 1.661 \times 10^{-27} \times 1.602 \times 10^{-19}}$$

$$= 9.69 \times 10^7 \, \text{eV} = 96.9 \, \text{MeV}.$$

8.2 (i) From (8.12)

$$m = E/c^2 + m_0 = 10^{12} \times 1.602 \times 10^{-19}/(2.998 \times 10^8)^2$$

$$+ 238 \times 1.661 \times 10^{-27} = 2.178 \times 10^{-24} \, \text{kg}.$$

(ii) From (8.8)

$$v = c\sqrt{1 - \frac{m_0^2}{m^2}}$$

$$= 2.998 \times 10^8 \sqrt{1 - \left(\frac{238 \times 1.661 \times 10^{-27}}{2.178 \times 10^{-24}}\right)^2}$$

$$= 2.948 \times 10^8 \, \text{m s}^{-1}.$$

(iii) The time for one circuit of the LHC is

$$t = \frac{2.7 \times 10^4}{2.948 \times 10^8} = 9.159 \times 10^{-5} \, \text{s}.$$

Chapter 9

Exercises 9

9.1 The natural line-width is

$$\Delta\nu_N = \frac{1}{4\pi \times 1.625 \times 10^{-8}} = 4.90 \times 10^6 \, \text{Hz}.$$

9.2 The root-mean-square speed of the mercury atoms is

$$\overline{V^2}^{1/2} = \sqrt{\frac{3 \times 1.381 \times 10^{-23} \times 700}{201 \times 1.661 \times 10^{-27}}} = 294.7\,\mathrm{m\,s^{-1}}.$$

The frequency of the spectral line is

$$\nu = \frac{2.998 \times 10^8}{4.358 \times 10^{-7}} = 6.879 \times 10^{14}\,\mathrm{Hz}.$$

Inserting $2\overline{V^2}^{1/2}$ and ν in (9.3) gives

$$\Delta\nu_D = 6.879 \times 10^{14} \times \frac{589.4}{2.998 \times 10^8} = 1.352 \times 10^9\,\mathrm{Hz}.$$

9.3 The recoil change of frequency is

$$\Delta\nu_R = \frac{6.626 \times 10^{-34} \times (6.879 \times 10^{14})^2}{2 \times 202 \times 1.661 \times 10^{-27} \times (2.998 \times 10^8)^2} = 5,199\,\mathrm{Hz}.$$

9.4 The frequency is

$$\nu = \frac{E}{h} = \frac{129 \times 1.602 \times 10^{-16}}{6.626 \times 10^{-34}} = 3.12 \times 10^{19}\,\mathrm{Hz}.$$

9.5 The natural line-width is

$$\Delta\nu_N = \frac{1}{4\pi \times 1.01 \times 10^{=10}} = 7.89 \times 10^9\,\mathrm{Hz}.$$

9.6 From (9.3)

$$2 \times \frac{3.48 \times 10^{18} \times \overline{V^2}^{1/2}}{2.998 \times 10^8} = 10^{14}$$

giving

$$\overline{V^2}^{1/2} = \frac{10^{14} \times 2.998 \times 10^8}{2 \times 3.48 \times 10^{18}} = 4,307\,\mathrm{m\,s^{-1}}.$$

Now, from (9.5) the required temperature is

$$T = \frac{\overline{V^2}m}{3k} = \frac{(4,307)^2 \times 57 \times 1.661 \times 10^{-27}}{3 \times 1.381 \times 10^{-23}} = 4.24 \times 10^4\,\mathrm{K}.$$

9.7 From Exercise 9.4 the frequency of the photon is 3.12×10^{19} Hz. Hence the reduction of frequency due to recoil is

$$\Delta \nu_R = \frac{6.626 \times 10^{-34} \times (3.12 \times 10^{19})^2}{2 \times 191 \times 1.661 \times 10^{-27} \times (2.998 \times 10^8)^2}$$

$$= 1.13 \times 10^{13} \text{ Hz}.$$

9.8 The change of frequency due to the isomer shift is

$$\Delta \nu = \frac{\Delta E}{h} = \frac{10^{-7} \times 1.602 \times 10^{-19}}{6.626 \times 10^{-34}} = 2.418 \times 10^7 \text{ Hz}.$$

The corresponding carriage speed is

$$V = \frac{c\Delta\nu}{\nu} = \frac{2.998 \times 10^8 \times 2.418 \times 10^7}{3.48 \times 10^{18}}$$

$$= 2.08 \times 10^{-3} \text{ m s}^{-1} = 2.08 \text{ mm s}^{-1}.$$

9.9 The speed of the emitter is

$$v = \frac{0.01}{3600} \text{ m s}^{-1}.$$

From (9.3) and (9.9)

$$l = \frac{\Delta\nu}{\nu} \frac{c^2}{g} = \frac{vc}{g} = \frac{0.01 \times 2.998 \times 10^8}{3600 \times 9.81} = 84.89 \text{ m}.$$

Problem 9

9.1 First we transform the velocities into energy difference of the maximum absorption positions from the energy E_0. From (9.3)

$$\delta E = h\delta\nu = h\nu V/c = \alpha V.$$

The conversion factor α, to give energy in electron volts, expressing the speed of light in mm s^{-1}. is

$$\alpha = \frac{h\nu}{c} = \frac{6.626 \times 10^{-34} \times 3.48 \times 10^{18}}{2.998 \times 10^{11} \times 1.602 \times 10^{-19}} = 4.801 \times 10^{-8} \text{ eV mm}^{-1} \text{ s}.$$

This gives energy differences, in units of 10^{-7} eV

$$-4.85 \quad -2.93 \quad -0.59 \quad 1.25 \quad 3.49 \quad 5.86.$$

From Figure 9.8 we can now express the energies of maximum absorption in terms of differences in the energy levels.

$$E(3/2, -3/2) - E(1/2, -1/2) = E_0 - 4.85 \quad \text{(i)}$$
$$E(3/2, -1/2) - E(1/2, -1/2) = E_0 - 2.93 \quad \text{(ii)}$$
$$E(3/2, 1/2) - E(1/2, -1/2) = E_0 - 0.59 \quad \text{(iii)}$$
$$E(3/2, -1/2) - E(1/2, 1/2) = E_0 + 1.25 \quad \text{(iv)}$$
$$E(3/2, 1/2) - E(1/2, 1/2) = E_0 + 3.49 \quad \text{(v)}$$
$$E(3/2, 3/2) - E(1/2, 1/2) = E_0 + 5.86 \quad \text{(vi)}$$

Subtracting (ii) from (iv)

$$E(1/2, -1/2) - E(1/2, 1/2) = 4.18 \quad \text{giving } E(1/2, -1/2) = 4.18.$$

Subtracting (iii) from (v)

$$E(1/2, -1/2) - E(1/2, 1/2) = 4.08 \quad \text{giving } E(1/2, -1/2) = 4.08.$$

We take the average to give, in units of 10^{-7} eV, $E(1/2, -1/2) = 4.13$.

From (i) $E(3/2, -3/2) = E_0 - 4.85 + E(1/2, -1/2) = E_0 - 0.72$.
From (ii) $E(3/2, -1/2) = E_0 - 2.93 + E(1/2, -1/2) = E_0 + 1.20$.
From (iv) $E(3/2, -1/2) = E_0 + 1.25 + E(1/2, 1/2) = E_0 + 1.25$.
We take an average value $E(3/2, -1/2) = E_0 + 1.225$.
From (iii) $E(3/2, 1/2) = E_0 - 0.59 + E(1/2, -1/2) = E_0 + 3.54$.
From (v) $E(3/2, 1/2) = E_0 + 3.49 + E(1/2, 1/2) = E_0 + 3.49$.

We take an average value $E(3/2, 1/2) = E_0 + 3.515$.
From (vi) $E(3/2, 3/2) - E(1/2, 1/2) = E(3/2, 3/2) = E_0 + 5.86$.

Listing the energies

$E(1/2, 1/2) = 0$ $E(1/2, -1/2) = 4.13 \times 10^{-7}$ eV
$E(3/2, -3/2) = E_0 - 7.2 \times 10^{-8}$ eV $E(3/2, -1/2) = E_0 + 1.225 \times 10^{-7}$ eV
$E(3/2, 1/2) = E_0 + 3.515 \times 10^{-7}$ eV $E(3/2, 3/2) = E_0 + 5.86 \times 10^{-7}$ eV

Appendix AI

Examples AI

AI.1 The expansion is $1 + \frac{6!}{5!1!}x + \frac{6!}{4!2!}x^2 + \frac{6!}{3!3!}x^3 + \frac{6!}{2!4!}x^4 + \frac{6!}{1!5!}x^5 + \frac{6!}{0!6!}x^6$

$= 1 + 6x + 15x^2 + 20x^3 + 15x^4 + 6x^5 + x^6$ (note: $0! = 1$).

AI.2

$$\frac{(1+2x)^3}{(0.5+x)^{4/3}} = \frac{(1+2x)^3}{0.5^{4/3}(1+2x)^{4/3}} = 2.5198(1+2x)^{5/3}$$

$$\approx 2.5198(1 + 10x/3) = 2.5198 + 8.3995x.$$

For $x = 0.01$ the true value is 2.6044. The approximation is 2.6038.

For $x = 0.005$ the true value is 2.5620. The approximation is 2.5618.

Index